Mining Methods: Engineering Fundamentals

Mining Methods: Engineering Fundamentals

Diana Bates

SYRAWOOD
PUBLISHING HOUSE

New York

Published by Syrawood Publishing House,
750 Third Avenue, 9th Floor,
New York, NY 10017, USA
www.syrawoodpublishinghouse.com

Mining Methods: Engineering Fundamentals
Diana Bates

International Standard Book Number: 978-1-68286-816-4 (Hardback)

This book contains information obtained from authentic and highly regarded sources. All chapters are published with permission under the Creative Commons Attribution Share Alike License or equivalent. A wide variety of references are listed. Permissions and sources are indicated; for detailed attributions, please refer to the permissions page. Reasonable efforts have been made to publish reliable data and information, but the authors, editors and publisher cannot assume any responsibility for the validity of all materials or the consequences of their use.

Trademark Notice: Registered trademark of products or corporate names are used only for explanation and identification without intent to infringe.

Cataloging-in-Publication Data

Mining methods : engineering fundamentals / Diana Bates.
    p. cm.
Includes bibliographical references and index.
ISBN 978-1-68286-816-4
1. Mining engineering. 2. Mines and mineral resources. 3. Engineering. I. Bates, Diana.
TN153 .M56 2019
622--dc23

TABLE OF CONTENTS

Contents

Permissions

Index

PREFACE

The extraction of valuable geological materials and minerals from the earth is referred to as mining. These are extracted as ores from an orebody, vein, lode, reef, etc. which are then processed with mechanical and chemical means to recover metals, gemstones, limestone, coal, etc. The extraction of petroleum and natural gas is also encompassed within this field. The techniques used in mining fall under the broad classification of surface mining and underground or sub-surface mining. Heavy machinery is used in mining at various levels. This book is a compilation of chapters that discuss the most vital concepts and methods in the field of mining. Different approaches, evaluations and methodologies have been included herein. Coherent flow of topics, student-friendly language and extensive use of examples make this book an invaluable source of knowledge.

To facilitate a deeper understanding of the contents of this book a short introduction of every chapter is written below:

Chapter 1- The extraction of minerals and geological materials usually from a lode, vein, seam, reef, etc. is called mining. Some of the ores recovered through mining include coal, metals, gemstones, chalk, rock salt, limestone, etc. This chapter has been carefully written to provide an easy understanding of mining, its different types such as coal mining, gold mining, silver mining, etc. and mining engineering.

Chapter 2- Mining can be classified into two common excavation types, namely surface and underground mining. The aim of this chapter is to explore these forms of mining and their varied aspects such as open-pit and cut mining, strip mining, hard-rock underground mining, soft-rock underground mining, etc. These topics are crucial for a complete understanding of the field.

Chapter 3- Heavy machinery is employed for exploring and developing sites, breaking and removing rocks, removing and stockpiling overburden or for processing the ore. Some of the mining equipment commonly used are power shovel, LHD, dragline excavator, gold dredge, crusher, etc. which are vital for drilling and blasting rocks. The topics elaborated in this chapter on such mining equipment and the processes of drilling and blasting, rock blasting, etc. will add valuable insights into the technical aspects of mining.

Chapter 4- In order to completely understand mining and mining processes, it is necessary to understand the mechanical infrastructure intrinsic to transportation systems in mines. Such transportation systems include conveyor system, trams, headframe, loader, winding engine, etc. The following chapter elucidates these tools and machinery associated with transportation in mines.

Chapter 5- Safety has been a major concern in the mining industry. Mining ventilation, gas ignition, cave-ins and rock falls, heat strokes, etc. are some of the hazards in mines. The varied ways of ensuring safety in mines, such as through the implementation of personal protective equipment, self-contained self-rescue device, good mine ventilation systems, safety lamp, etc. have been elaborately discussed in this chapter.

I would like to share the credit of this book with my editorial team who worked tirelessly on this book. I owe the completion of this book to the never-ending support of my family, who supported me throughout the project.

Diana Bates

Chapter 1

Introduction to Mining

The extraction of minerals and geological materials usually from a lode, vein, seam, reef, etc. is called mining. Some of the ores recovered through mining include coal, metals, gemstones, chalk, rock salt, limestone, etc. This chapter has been carefully written to provide an easy understanding of mining, its different types such as coal mining, gold mining, silver mining, etc. and mining engineering.

Mining is the extraction of valuable minerals or other geological materials from the earth from an orebody, lode, vein, seam, or reef, which forms the mineralized package of economic interest to the miner.

Ores recovered by mining include metals, coal, oil shale, gemstones, limestone, dimension stone, rock salt, potash, gravel, and clay. Mining is required to obtain any material that cannot be grown through agricultural processes, or created artificially in a laboratory or factory. Mining in a wider sense includes extraction of any non-renewable resource such as petroleum, natural gas, or even water.

Figure: Surface coal mining

Mining of stone and metal has been done since pre-historic times. Modern mining processes involve prospecting for ore bodies, analysis of the profit potential of a proposed mine, extraction of the desired materials, and final reclamation of the land after the mine is closed.

The nature of mining processes creates a potential negative impact on the environment both during the mining operations and for years after the mine is closed. This impact has led most of the world's nations to adopt regulations designed to moderate the negative effects of mining operations. Safety has long been a concern as well, and modern practices have improved safety in mines significantly.

Mine Development and Lifecycle

The process of mining from discovery of an ore body through extraction of minerals and finally to returning the land to its natural state consists of several distinct steps. The first is discovery of the ore body, which is carried out through prospecting or exploration to find and then define the extent, location and value of the ore body. This leads to a mathematical resource estimation to estimate the size and grade of the deposit.

This estimation is used to conduct a pre-feasibility study to determine the theoretical economics of the ore deposit. This identifies, early on, whether further investment in estimation and engineering studies is warranted and identifies key risks and areas for further work. The next step is to conduct a feasibility study to evaluate the financial viability, the technical and financial risks, and the robustness of the project.

This is when the mining company makes the decision whether to develop the mine or to walk away from the project. This includes mine planning to evaluate the economically recoverable portion of the deposit, the metallurgy and ore recoverability, marketability and payability of the ore concentrates, engineering concerns, milling and infrastructure costs, finance and equity requirements, and an analysis of the proposed mine from the initial excavation all the way through to reclamation. The proportion of a deposit that is economically recoverable is dependent on the enrichment factor of the ore in the area.

To gain access to the mineral deposit within an area it is often necessary to mine through or remove waste material which is not of immediate interest to the miner. The total movement of ore and waste constitutes the mining process. Often more waste than ore is mined during the life of a mine, depending on the nature and location of the ore body. Waste removal and placement is a major cost to the mining operator, so a detailed characterization of the waste material forms an essential part of the geological exploration program for a mining operation.

Once the analysis determines a given ore body is worth recovering, development begins to create access to the ore body. The mine buildings and processing plants are built, and any necessary equipment is obtained. The operation of the mine to recover the ore begins and continues as long as the company operating the mine finds it economical to do so. Once all the ore that the mine can produce profitably is recovered, reclamation begins to make the land used by the mine suitable for future use.

Mining Techniques

Mining techniques can be divided into two common excavation types: surface mining and sub-surface (underground) mining. Today, surface mining is much more common, and produces, for example, 85% of minerals (excluding petroleum and natural gas) in the United States, including 98% of metallic ores.

Targets are divided into two general categories of materials: *placer deposits*, consisting of valuable minerals contained within river gravels, beach sands, and other unconsolidated materials; and *lode deposits*, where valuable minerals are found in veins, in layers, or in mineral grains generally distributed throughout a mass of actual rock. Both types of ore deposit, placer or lode, are mined by both surface and underground methods.

Some mining, including much of the rare earth elements and uranium mining, is done by less-common methods, such as in-situ leaching: this technique involves digging neither at the surface nor underground. The extraction of target minerals by this technique requires that they be soluble, e.g., potash, potassium chloride, sodium chloride, sodium sulfate, which dissolve in water. Some minerals, such as copper minerals and uranium oxide, require acid or carbonate solutions to dissolve.

Surface Mining

Surface mining is done by removing (stripping) surface vegetation, dirt, and, if necessary, layers of bedrock in order to reach buried ore deposits. Techniques of surface mining include: open-pit mining, which is the recovery of materials from an open pit in the ground, quarrying or gathering building materials from an open-pit mine; strip mining, which consists of stripping surface layers off to reveal ore/ seams underneath; and mountaintop removal, commonly associated with coal mining, which involves taking the top of a mountain off to reach ore deposits at depth. Most (but not all) placer deposits, because of their shallowly buried nature, are mined by surface methods. Finally, landfill mining involves sites where landfills are excavated and processed.

Figure: Garzweiler surface mine, Germany

Open-pit Mining

Open-pit mining, or open-cast mining is a surface mining technique of extracting rock or minerals from the earth by their removal from an open pit or borrow.

This form of mining differs from extractive methods that require tunneling into the earth, such as long wall mining. Open-pit mines are used when deposits of commercially useful minerals or rocks are found near the surface; that is, where the overburden (surface material covering the valuable deposit) is relatively thin or the material of interest is structurally unsuitable for tunneling (as would be the case for sand, cinder, and gravel). For minerals that occur deep below the surface—where the overburden is thick or the mineral occurs as veins in hard rock—underground mining methods extract the valued material.

Open-pit mines that produce building materials and dimension stone are commonly referred to as "quarries."

Open-pit mines are typically enlarged until either the mineral resource is exhausted, or an increasing ratio of overburden to ore makes further mining uneconomic. When this occurs, the exhausted mines are sometimes converted to landfills for disposal of solid wastes. However, some form of water control is usually required to keep the mine pit from becoming a lake, if the mine is situated in a climate of considerable precipitation or if any layers of the pit forming the mine border productive aquifers.

Open-cast mines are dug on benches, which describe vertical levels of the hole. These benches are usually on four to sixty meter intervals, depending on the size of the machinery that is being used. Many quarries do not use benches, as they are usually shallow.

Most walls of the pit are generally dug on an angle less than vertical, to prevent and minimize damage and danger from rock falls. This depends on how weathered the rocks are, and the type of rock, and also how many structural weaknesses occur within the rocks, such as a faults, shears, joints orfoliations.

Figure: The angled and stepped sides of the Sunrise Dam Gold Mine

The walls are stepped. The inclined section of the wall is known as the batter, and the flat part of the step is known as the bench or berm. The steps in the walls help prevent rock falls continuing down the entire face of the wall. In some instances additional ground support is required and rock bolts, cable bolts and shotcrete are used. De-watering bores may be used to relieve water pressure by drilling horizontally into the wall, which is often enough to cause failures in the wall by itself.

A haul road is usually situated at the side of the pit, forming a ramp up which trucks can drive, carrying ore and waste rock.

Waste rock is piled up at the surface, near the edge of the open pit. This is known as the waste dump. The waste dump is also tiered and stepped, to minimize degradation.

Ore which has been processed is known as tailings, and is generally a slurry. This is pumped to a tailings dam or settling pond, where the water evaporates. Tailings dams can often be toxic due to the presence of unextracted sulfide minerals, some forms of toxic minerals in the gangue, and often cyanide which is used to treat gold ore via the cyanide leach process. This toxicity can harm the surrounding environment.

Gold is generally extracted in open-pit mines at 1 to 2 ppm (parts per million) but in certain cases, 0.75 ppm gold is economical. This was achieved by bulk heap leaching at the Peak Hill mine in western New South Wales, near Dubbo, Australia.

Nickel, generally as laterite, is extracted via open-pit down to 0.2%. Copper is extracted at grades as low as 0.15% to 0.2%, generally in massive open-pit mines in Chile, where the size of the resources and favorable metallurgy allows economies of scale.

Materials typically extracted from open-pit mines include:

- Bitumen
- Clay
- Coal
- Copper
- Coquina
- Diamonds
- Gravel and stone (stone refers to bedrock, while gravel is unconsolidated material)
- Granite
- Gritstone

- Gypsum

- Limestone

- Marble

- Metal ores, such as Copper, Iron, Gold, Silver and Molybdenum

- Uranium

- Phosphate

- Underground Mining

Sub-surface mining consists of digging tunnels or shafts into the earth to reach buried ore deposits. Ore, for processing, and waste rock, for disposal, are brought to the surface through the tunnels and shafts. Sub-surface mining can be classified by the type of access shafts used, the extraction method or the technique used to reach the mineral deposit. Drift mining utilizes horizontal access tunnels, slope mining uses diagonally sloping access shafts, and shaft mining utilizes vertical access shafts. Mining in hard and soft rock formations require different techniques.

Figure. Mantrip used for transporting miners within an underground mine

Other methods include shrinkage stope mining, which is mining upward, creating a sloping underground room, long wall mining, which is grinding a long ore surface underground, and room and pillar mining, which is removing ore from rooms while leaving pillars in place to support the roof of the room. Room and pillar mining often leads to retreat mining, in which supporting pillars are removed as miners retreat, allowing the room to cave in, thereby loosening more ore. Additional sub-surface mining methods include hard rock mining, which is mining of hard rock (igneous, metamorphic or sedimentary) materials, bore hole mining, drift and fill mining, long hole slope mining, sub level caving, and block caving.

Machines

Heavy machinery is used in mining to explore and develop sites, to remove and stockpile overburden, to break and remove rocks of various hardness and toughness, to process the ore, and to carry out reclamation projects after the mine is closed. Bulldozers, drills, explosives and trucks are all necessary for excavating the land. In the case of placer mining, unconsolidated gravel, or alluvium, is fed into machinery consisting of a hopper and a shaking screen or trommel which frees the desired minerals from the waste gravel. The minerals are then concentrated using sluices or jigs.

Large drills are used to sink shafts, excavate stopes, and obtain samples for analysis. Trams are used to transport miners, minerals and waste. Lifts carry miners into and out of mines, and move rock and ore out, and machinery in and out, of underground mines. Huge trucks, shovels and cranes are employed in surface mining to move large quantities of overburden and ore. Processing plants utilize large crushers, mills, reactors, roasters and other equipment to consolidate the mineral-rich material and extract the desired compounds and metals from the ore.

Figure: The Bagger 288 is a bucket-wheel excavator used in strip mining.
It is also the largest land vehicle of all time.

Processing

Once the mineral is extracted, it is often then processed. The science of extractive metallurgy is a specialized area in the science of metallurgy that studies the extraction of valuable metals from their ores, especially through chemical or mechanical means.

Mineral processing (or mineral dressing) is a specialized area in the science of metallurgy that studies the mechanical means of crushing, grinding, and washing that enable the separation (extractive metallurgy) of valuable metals or minerals from their gangue (waste material). Processing of placer ore material consists of gravity-dependent methods of separation, such as sluice boxes. Only minor shaking or washing may be necessary to disaggregate (unclump) the sands or gravels before processing. Processing of ore from a lode mine, whether it is a surface or subsurface mine, requires that the rock ore be crushed and pulverized before extraction of the valuable minerals begins. After lode ore is crushed, recovery of the valuable minerals is done by one, or a combination of several, mechanical and chemical techniques.

Since most metals are present in ores as oxides or sulfides, the metal needs to be reduced to its metallic form. This can be accomplished through chemical means such as smelting or through electrolytic reduction, as in the case of aluminium. Geometallurgy combines the geologic sciences with extractive metallurgy and mining.

Mining Industry

Mining exists in many countries. London is known as the capital of global "mining houses" such as Rio Tinto Group, BHP Billiton, and Anglo American PLC. The US mining industry is also large, but it is dominated by the coal and other nonmetal minerals (e.g., rock and sand), and various regulations have worked to reduce the significance of mining in the United States. In 2007 the totalmarket capitalization of mining companies was reported at US$962 billion, which compares to a total global market cap of publicly traded companies of about US$50 trillion in 2007. In 2002, Chile and Peru were reportedly the major mining countries of South America. The mineral industry of Africa includes the mining of various minerals; it produces relatively little of the industrial metals copper, lead, and zinc, but according to one estimate has as a percent of world reserves 40% of gold, 60% of cobalt, and 90% of the world's platinum group metals. Mining in India is a significant part of that country's economy. In the developed world, mining in Australia, with BHP Billiton founded and headquartered in the country, and mining in Canada are particularly significant. For rare earth minerals mining, China reportedly controlled 95% of production in 2013.

Mining operations can be grouped into five major categories in terms of their respective resources. These are oil and gas extraction, coal mining, metal ore mining, nonmetallic mineral mining and quarrying, and mining support activities. Of all of these categories, oil and gas extraction remains one of the largest in terms of its global economic importance. Prospecting potential mining sites, a vital area of concern for the mining industry, is now done using sophisticated new technologies such as seismic prospecting and remote-sensing satellites. Mining is heavily affected by the prices of the commodity minerals, which are often volatile. The 2000s commodities boom ("commodities supercycle") increased the prices of commodities, driving aggressive mining. In addition, the price of gold increased dramatically in the 2000s, which increased gold mining; for example, one study found that conversion of forest in the Amazon increased six-fold from the period 2003–2006 (292 ha/yr) to the period 2006–2009 (1,915 ha/yr), largely due to artisanal mining.

Safety

Safety has long been a concern in the mining business especially in sub-surface mining. The Courrières mine disaster, Europe's worst mining accident, involved the death of 1,099 miners in Northern France on March 10, 1906. This disaster was surpassed only by the Benxihu Colliery accident in China on April 26, 1942, which killed 1,549 miners. While mining today is substantially safer than it was in previous decades, mining accidents still occur. Government figures indicate that 5,000 Chinese miners die in accidents each year, while other reports have suggested a figure as high as 20,000. Mining accidents continue worldwide, including accidents causing dozens of fatalities at a time such as the 2007 Ulyanovskaya Mine disaster in Russia, the 2009 Heilongjiang mine explosion in China, and the 2010 Upper Big Branch Mine disaster in the United States.

Mining ventilation is a significant safety concern for many miners. Poor ventilation inside sub-surface mines causes exposure to harmful gases, heat, and dust, which can cause illness, injury, and death. The concentration of methane and other air-borne contaminants underground can generally be controlled by dilution (ventilation), capture before entering the host air stream (methane drainage), or isolation (seals and stoppings). Rock dusts, including coal dust and silicon dust, can cause long-term lung problems including silicosis, asbestosis, and pneumoconiosis (also known as miners lung or black lungdisease). A ventilation system is set up to force a stream of air through the working areas of the mine. The air circulation necessary for effective ventilation of a mine is generated by one or more large mine fans, usually located above ground. Air flows in one direction only, making circuits through the mine such that each main work area constantly receives a supply of fresh air. Watering down in coal mines also helps to keep dust levels down: by spraying the machine with water and filtering the dust-laden water with a scrubber fan, miners can successfully trap the dust.

Gases in mines can poison the workers or displace the oxygen in the mine, causing asphyxiation. For this reason, the U.S. Mine Safety and Health Administration requires that groups of miners in the United States carry gas detection equipment that can detect common gases, such as CO, O_2, H_2S, CH_4, as well as calculate % Lower Explosive Limit. Regulation requires that all production stop if there is a concentration of 1.4% of flammable gas present. Additionally, further regulation is being requested for more gas detection as newer technology such as nanotechnology is introduced.

Ignited methane gas is a common source of explosions in coal mines, which in turn can initiate more extensive coal dust explosions. For this reason, rock dusts such as limestone dust are spread throughout coal mines to diminish the chances of coal dust explosions as well as to limit the extent of potential explosions, in a process known as rock dusting. Coal dust explosions can also begin independently of methane gas explosions. Frictional heat and sparks generated by mining equipment can ignite both methane gas and coal dust. For this reason, water is often used to cool rock-cutting sites.

Miners utilize equipment strong enough to break through extremely hard layers of the Earth's crust. This equipment, combined with the closed work space in which underground miners work, can cause hearing loss. For example, a roof bolter (commonly used by mine roof bolter operators) can reach sound power levels of up to 115 dB. Combined with the reverberant effects of underground mines, a miner without proper hearing protection is at a high risk for hearing loss. By age 50, nearly 90% of U.S. coal miners have some hearing loss, compared to only 10% among workers not exposed to loud noises. Roof bolters are among the loudest machines, but auger miners, bulldozers, continuous mining machines, front end loaders, and shuttle cars and trucks are also among those machines most responsible for excessive noise in mine work.

Since mining entails removing dirt and rock from its natural location, thereby creating large empty pits, rooms, and tunnels, cave-ins as well as ground and rock falls are a major concern within mines. Modern techniques for timbering and bracing walls and ceilings within sub-surface mines have reduced the number of fatalities due to cave-ins, but ground falls continue to represent up to 50% of mining fatalities. Even in cases where mine collapses are not instantly fatal, they can trap mine workers deep underground. Cases such as these often lead to high-profile rescue efforts, such as when 33 Chilean miners were trapped deep underground for 69 days in 2010.

High temperatures and humidity may result in heat-related illnesses, including heat stroke, which can be fatal. The presence of heavy equipment in confined spaces also poses a risk to miners. To improve the safety of mine workers, modern mines use automation and remote operation including, for example, such equipment as automated loaders and remotely operated rockbreakers. However, despite modern improvements to safety practices, mining remains a dangerous occupation throughout the world.

Effects of Mining

Environmental Effects

Environmental issues can include erosion, formation of sinkholes, loss of biodiversity, and contamination of soil, groundwater and surface water by chemicals from mining processes. In some cases, additional forest logging is done in the vicinity of mines to create space for the storage of the created debris and soil. Contamination resulting from leakage of chemicals can also affect the health of the local population if not properly controlled. Extreme examples of pollution from mining activities include coal fires, which can last for years or even decades, producing massive amounts of environmental damage.

Mining companies in most countries are required to follow stringent environmental and rehabilitation codes in order to minimize environmental impact and avoid impacting human health. These codes and regulations all require the common steps of environmental impact assessment, development of environmental management plans, mine closure planning (which must be done before the start of mining operations), and environmental monitoring during operation and after closure. However, in some areas, particularly in the developing world, government regulations may not be well enforced.

Waste

Ore mills generate large amounts of waste, called tailings. For example, 99 tons of waste are generated per ton of copper, with even higher ratios in gold mining. These tailings can be toxic. Tailings, which are usually produced as a slurry, are most commonly dumped into ponds made from naturally existing valleys. These ponds are secured by impoundments (dams or embankment dams). In 2000 it was estimated that 3,500 tailings impoundments existed, and that every year, 2 to 5 major failures and 35 minor failures occurred; for example, in the Marcopper mining disaster at least 2 million tons of tailings were released into a local river. Subaqueous tailings disposal is another option. The mining industry has argued that submarine tailings disposal (STD), which disposes of tailings in the sea, is ideal because it avoids the risks of tailings ponds; although the practice is illegal in the United States and Canada, it is used in the developing world.

The waste is classified as either sterile or mineralised, with acid generating potential, and the movement and storage of this material forms a major part of the mine planning process. When the mineralised package is determined by an economic cut-off, the near-grade mineralised waste is usually dumped separately with view to later treatment should market conditions change and it becomes economically viable. Civil engineering design parameters are used in the design of the waste dumps, and special conditions apply to high-rainfall areas and to seismically active areas. Waste dump designs must meet all regulatory requirements of the country in whose jurisdiction the mine is located. It is also common practice to rehabilitate dumps to an internationally acceptable standard, which in some cases means that higher standards than the local regulatory standard are applied.

Open-Pit Mining

After mining finishes, the mine area must undergo rehabilitation. Waste dumps are contoured to flatten them out, to further stabilise them. If the ore contains sulfides it is usually covered with a layer of clay to prevent access of rain and oxygen from the air, which can oxidise the sulfides to produce sulfuric acid, a phenomenon known as acid mine drainage. This is then generally covered with soil, and vegetation is planted to help consolidate the material. Eventually this layer will erode, but it is generally hoped that the rate of leaching or acid will be slowed by the cover such that the environment can handle the load of acid and associated heavy metals. There are no long term studies on the success of these covers due to the relatively short time in which large scale open pit mining has existed. It may take hundreds to thousands of years for some waste dumps to become "acid neutral" and stop leaching to the environment. The dumps are usually fenced off to prevent livestock denuding them of vegetation. The open pit is then surrounded with afence, to prevent access, and it generally eventually fills up with ground water. In arid areas it may not fill due to deep groundwater levels.

Figure: Opencut coal mine loadout station and reclaimed land at the North Antelope Rochelle coal mine in Wyoming, United States.

Figure: An open-pit sulfur mine at Tarnobrzeg, Poland undergoing land rehabilitation

Metal Reserves and Recycling

During the twentieth century, the variety of metals used in society grew rapidly. Today, the development of major nations such as China and India and advances in technologies are fueling an ever greater demand. The result is that metal mining activities are expanding and more and more of the world's metal stocks are above ground in use rather than below ground as unused reserves. An example is the in-use stock of copper. Between 1932 and 1999, copper in use in the USA rose from 73 kilograms (161 lb) to 238 kilograms (525 lb) per person.

95% of the energy used to make aluminum from bauxite ore is saved by using recycled material. However, levels of metals recycling are generally low. In 2010, the International Resource Panel, hosted by the United Nations Environment Programme (UNEP), published reports on metal stocks that exist within society and their recycling rates.

The report's authors observed that the metal stocks in society can serve as huge mines above ground. However, they warned that the recycling rates of some rare metals used in applications such as mobile phones, battery packs for hybrid cars, and fuel cells are so low that unless future end-of-life recycling rates are dramatically stepped up these critical metals will become unavailable for use in modern technology.

Common Types of Minerals Mining

Coal Mining

Coal miners use giant machines to remove coal from the ground. They use two methods:

surface or underground mining. Many U.S. coal beds are very near the ground's surface, and about two-thirds of coal production comes from surface mines. Modern mining methods allow us to easily reach most of our coal reserves. Due to growth in surface mining and improved mining technology, the amount of coal produced by one miner in one hour has more than tripled since 1978.

Surface mining is used to produce most of the coal in the U.S. because it is less expensive than underground mining. Surface mining can be used when the coal is buried less than 200 feet underground. In surface mining, giant machines remove the top-soil and layers of rock to expose large beds of coal. Once the mining is finished, the dirt and rock are returned to the pit, the topsoil is replaced, and the area is replanted. The land can then be used for croplands, wildlife habitats, recreation, or offices or stores.

Underground mining, sometimes called deep mining, is used when the coal is buried several hundred feet below the surface. Some underground mines are 1,000 feet deep. To remove coal in these underground mines, miners ride elevators down deep mine shafts where they run machines that dig out the coal.

Processing the Coal

After coal comes out of the ground, it typically goes on a conveyor belt to a preparation plant that is located at the mining site. The plant cleans and processes coal to remove dirt, rock, ash, sulfur, and other unwanted materials, increasing the heating value of the coal.

Gold Mining

The major ores of gold contain gold in its native form and are both exogenetic (formed at the Earth's surface) and endogenetic (formed within the Earth). The best-known of the exogenetic ores is alluvial gold. Alluvial gold refers to gold found in riverbeds, streambeds, and floodplains. It is invariably elemental gold and usually made up of very fine particles. Alluvial gold deposits are formed through the weathering actions of wind, rain, and temperature change on rocks containing gold. They were the type most commonly mined in antiquity. Exogenetic gold can also exist as oxidized ore bodies that have formed under a process called secondary enrichment, in which other metallic elements and sulfides are gradually leached away, leaving behind gold and insoluble oxide minerals as surface deposits.

Endogenetic gold ores include vein and lode deposits of elemental gold in quartzite or mixtures of quartzite and various iron sulfide minerals, particularly pyrite (FeS_2) and pyrrhotite ($Fe_{1-x}S$). When present in sulfide ore bodies, the gold, although still elemental in form, is so finely disseminated that concentration by methods such as those applied to alluvial gold is impossible.

Native gold is the most common mineral of gold, accounting for about 80 percent of the metal in the Earth's crust. It occasionally is found as nuggets as large as 12 millimetres (0.5 inch) in diameter, and on rare occasions nuggets of native gold weighing up to 50 kilograms are found—the largest having weighed 92 kilograms. Native gold invariably contains about 0.1 to 4 percent silver. Electrum is a gold-silver alloy containing 20 to 45 percent silver. It varies from pale yellow to silver white in colour and is usually associated with silver sulfide mineral deposits.

Gold also forms minerals with the element tellurium; the most common of these are calaverite ($AuTe_2$) and sylvanite ($AuAgTe_4$). Other minerals of gold are sufficiently rare as to have little economic significance.

The Lifecycle of A Gold Mine

People in hard hats working underground is what often comes to mind when thinking about how gold is mined. Yet mining the ore is just one stage in a long and complex gold mining process. Long before any gold can be extracted, significant exploration and development needs to take place, both to determine, as accurately as possible, the size of the deposit as well as how to extract and process the ore efficiently, safely and responsibly. On average, it takes between 10-20 years before a gold mine is even ready to produce material that can be refined.

Gold Mine Exploration: 1 - 10 years

Gold mine exploration is challenging and complex. It requires significant time, financial resources and expertise in many disciplines – e.g. geography, geology, chemistry and engineering.

The likelihood of a discovery leading to a mine being developed is very low - less than 0.1% of prospected sites will lead to a productive mine. And only 10% of global gold deposits contain sufficient gold to justify further development.

Once basic facts about the local geology and potentially viable deposit are established, the ore body can be modelled and its feasibility assessed.

Gold Mine Development: 1 - 5 years

Gold mine development is the second stage of the gold mining process. It involves the planning and construction of the mine and associated infrastructure. Mining companies must obtain appropriate permits and licenses before they can begin construction. This will generally take several years, although this varies greatly depending on location.

Construction may not be confined to the mine itself. In addition to potential processing capacity, mining companies frequently construct local infrastructure and amenities to support both logistical and operational needs, as well as employee and community welfare. This development provides much long-term support for local communities, and one of the key ways gold supports economic development.

Gold Mining Operation: 10 - 30 years

The gold mining operation stage represents the productive life of a mine, during which ore is extracted and processed into gold. Processing involves transforming rock and ore into a metallic alloy of substantial purity – known as doré – typically containing between 60-90% gold.

During its life, a number of factors – such as the price of gold or input costs – will affect which areas of an ore body are deemed profitable (economic) to mine. In times of higher prices, mining low-grade ore will become profitable as the higher price offsets the increased cost of extracting and milling greater volumes. When the price is lower or costs rise, it might only prove profitable to extract and process higher-grade ores. Mine plans are regularly re-assessed as market conditions change and new technical information comes to light.

Gold Mine Decommissioning: 1 - 5 years

After a mine has ceased operations, possibly because the ore body is exhausted or the remaining deposit becomes unprofitable (uneconomic) to mine, work then focuses on its decommissioning, dismantling and rehabilitation of the land in which it was situated.

Gold mine closure is a complex undertaking. A mining company will also be required to monitor the mine site long after the mine has been closed.

Post-closure

Gold Mine Reclamation

Gold mining companies assume responsibility for the management of a site long after a mine has closed and been dismantled – typically for a period of five to ten years or more. Over this time, the land will be rehabilitated – cleansed and revegetated – and the mining company will work to ensure the gold mine reclamation and return to long-term environmental stability are successful.

The following images are from gold producer Newmont:

Before Reclamation

After Reclamation

Silver Mining

Silver is one of the most valued precious metals in the world. It is a key player in the world's monetary systems mainly being used to create bullion coins. Other than its use in currencies, silver also finds wide application in the creation of solar panels, jewelry, utensils, electrical conductors, water filtration, window coatings and mirrors among other things. Silver is also used the medical filed as disinfectants, in x-ray machines and other medical instruments.

Silver is a soft white metallic element represented by the symbol Ag and atomic number 47. The element is known to exhibit the highest reflectivity, thermal and electrical conductivity of any known metal. Silver is usually found in the crust of the earth either as a free element (native silver) or more commonly as an alloy of gold or other metallic elements.

It is not very abundant in native form and thus its purity is measured using a per mille measurement. In many places, silver ore is mined as a byproduct of gold, copper, zinc, or lead.

Characteristics of Silver

Silver has the same physical and chemical characteristics as its two group 11 neighbors in the periodic table: gold and copper.

Silver is a somewhat inert metal. This is on account of its filled 4d shell is not extremely powerful in protecting the electrostatic powers of attraction from the core to the outermost 5s electron. Among all the group 11 elements, silver has the most minimal first ionization energy, yet has higher second and third ionization energies than copper and gold.

It must be noted in spite of the above characteristics most silver compounds have more covalent character because of the high first ionization vitality (730.8 kJ/mol) and the small size of silver. Furthermore, silver's Pauling electronegativity of 1.93 is higher than that of lead (1.87), and its electron proclivity of 125.6 kJ/mol is much higher than that of hydrogen (72.8 kJ/mol) and very little not as much as that of oxygen (141.0 kJ/mol). Due to its full d-subshell, silver in its principle +1 oxidation state displays a few of properties of the transition metals appropriate from groups 4 to 10, forming rather unstable organometallic compounds.

47: Silver   2,8,18,
                     18,1

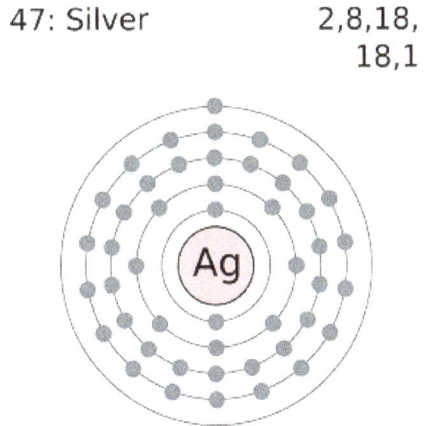

Silver just like gold and copper is soft malleable and extremely ductile. Silver characteristically crystallizes into a face-centered cubic lattice with mass coordination number 12, where just the single 5s electron is delocalized, just like the case of copper and gold. Metallic bonds in silver are inadequate with regards to a covalent character and are moderately weak. This helps explain the ductility and high malleability of silver.

Silver has very high electrical conductivity even when compared to copper and that is why it is widely applied in radio frequency engineering where high electrical conductivity is desired. Copper is widely used for most other application because the higher cost of silver often limits its use. Silver also has the highest thermal conductivity and the lowest resistance of any given metal.

Silver Mining Process

Silver is mined using a number of processes. One of the most common processes of extracting silver metal for the ore is the heap leach or cyanide process. The process is most popular with many miners because it is low cost, especially when processing low-grade ores.

To use the cyanide process the silver being in the ore should have smaller particles, should be able to react with the cyanide solution, the silver should be free from sulfide minerals and other foreign substances. The following are the major steps involved in silver mining using this method:

Preparation of Silver Ore

The first stage of mining silver involves the crushing the silver ore to about 1-1.5 in diameter so as to make the ore porous for the extraction process. Once the ore is crushed it is then mixed with lime (about 3-5 lb. per ton) to create a conducive alkaline conditions for the extractions. The ore is then stacked onto impermeable pads made of asphalt, rubber or plastic to ensure that there is a minimum loss of the silver cyanide solution once the extraction begins. Usually, the pads are arranged in a slanting position to allow for the drainage and the collection of the silver cyanide solution.

Curing with the Cyanide solution

The next step is to add a solution of sodium cyanide and water to the prepared silver ore. Usually, this may be done using a number of methods such as sprinkler systems or ponding method that involves seepage from capillaries, ditches or injection.

Recovering the Silver

Once the curing is done the silver cyanide solution has to be collected so that the silver can be extracted from the solution. The most commonly used method of recovering silver from the solution is by Crowe Precipitation. This method employs zinc dust to help precipitate silver from the solution. The silver collected is then filtered off then melted and then made into bullion bars.

The other method used to recover silver from the silver cyanide solution is the activated carbon absorption method in which the solution has to be pumped through towers with activated carbon to form a silver precipitate. The precipitate is then collected by filtration and melted before it is made into bars.

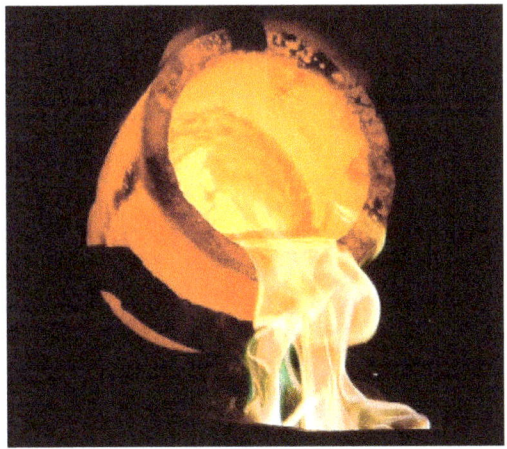

In many places, silver ore is often found in combination with other ores containing other commercially viable minerals such as copper, lead or gold. When silver is a byproduct of processing these other minerals then a different method has to be used to extract the silver ore. For example is silver is found in ore containing zinc a method known as the Parkes. Process is used to mine the minerals. When using this method the ore is first heated until it melts. When the ore is the cooled down a crust containing silver and zinc forms on the surface. The crust is then collected and then distilled to extract silver from the zinc.

When silver is found in ores containing copper then the electrolytic refining method is used to extract the silver. The ore is placed in an electrolyte solution with a cathode and an anode. Electricity is then passed through the solution forcing the silver to accumulate around the anode while copper attaches itself on the cathode. This is then collected and leached to remove impurities.

Mining Engineering

Mining Engineering is a specialized field of engineering and has gained acceptance from aspiring engineers, who want to do something different from core engineering that includes Computer Science, Mechanical or Chemical Engineering.

Mining Engineering is the science of extraction and processing of essential natural minerals from the environment. The discipline involves the practice, the theory, the science, the technology, and application of extracting and processing minerals from a naturally occurring environment. Mining Engineering also includes processing minerals for additional value.

Mining Industry is witnessing growing importance of recycling and reclamation. Mining Engineering involves using the latest extraction technology that allows treatment of secondary sources like industrial & domestic waste, waste from previously mined deposits, contaminated land, etc.

Mining engineers plan and direct the various engineering aspects of extracting minerals from the earth. They prepare initial plans for the type, size, location and construction of open pit or underground mines.

The sorts of things that a mining engineer oversees at a mine might include:

- Conduct investigations of mineral deposits and undertake evaluations in collaboration with geologists, other earth scientists and economists to determine whether the mineral deposits can be mined profitably.

- Prepare plans for mines, including tunnels and shafts for underground operations, and pits and haulage roads for open-cut operations, using computer-aided design packages.

- Prepare the layout of the mine development and the way the minerals are to be mined.

- Plan and coordinate the employment of mining staff and equipment with regard to efficiency, safety and environmental conditions.

- Consult with geologists and other engineers about the design, selection and provision of machines, facilities and systems for mining, as well as infrastructure such as access roads, water and power supplies.

- Operate computers to assist with calculations, prepare estimates on the cost of the operation and control expenditure when mines come into production.

- Oversee the construction of the mine and the installation of plant and equipment.

- Make sure that mining regulations are observed, including the proper use and care of explosives, and the correct ventilation to allow the removal of dust and gases.

- Conduct research aimed at improving efficiency and safety in mines.

- Establish first aid and emergency services facilities at the mines.

References

- Outcome-extraction-methods, wmopen-geology: lumenlearning.com, Retrieved 10 April 2018

- Coal-mining-and-processing: energytrendsinsider.com, Retrieved 19 June 2018

- Gold-processing-81536: britannica.com, Retrieved 24 May 2018

- Know-all-about-mining-engineering-why-and-how: engineering.careers360.com, Retrieved 24 June 2018

- Careers-booklet-miningeng: ausimm.com.au, Retrieved 14 July 2018

Chapter 2

Mining Techniques

Mining can be classified into two common excavation types, namely surface and underground mining. The aim of this chapter is to explore these forms of mining and their varied aspects such as open-pit and cut mining, strip mining, hard-rock underground mining, soft-rock underground mining, etc. These topics are crucial for a complete understanding of the field.

Surface Mining

Surface mining is a form of mining in which the soil and the rock covering the mineral deposits are removed. It is the other way of underground mining, in which the overlying rock is left behind, and the required mineral deposits are removed through shafts or tunnels.

Extraction of mineral or energy resources by operations exclusively involving personnel working on the surface without provision of manned underground operations is referred to as surface mining. While an opening may sometimes be constructed below the surface and limited underground development may occasionally be required, this type of mining is essentially surface-based. Surface mining can be classified into two groups on the basis of the method of extraction; mechanical extraction, or aqueous extraction.

Mechanical extraction methods employ mechanical processes in a dry environment to recover minerals, encompassing the specific mining methods of:

- Open pit mining

- Open cast mining

- Quarrying of dimension stone

- Highwall/auger mining

Open pit and open cast methods employ a conventional mining cycle of operations to extract minerals: rock breakage is usually accomplished by drilling and blasting for consolidated materials and by ripping or direct removal by excavators for unconsolidated soil and/or decomposed rock, followed by materials handling and transportation.

Dimension stone quarrying is quite similar to open pit mining, but rock breakage without blasting is almost exclusively employed to cut prismatic blocks or tabular slabs of rock. The high labor intensity and cost associated with cutting stone makes quarrying the most expensive surface mining method.

Highwall mining is a coal mining method for recovery of outcropped coal by mechanical excavation without removal of overburden. A continuous miner with single or multiple augers/cutting heads is operated underground and controlled remotely by crew located outside. Augering can be regarded as a supplementary method for open cast mining in cases when coal seams in the highwall would otherwise remain unmined (unless recoverable by underground methods) or when rugged terrain would preclude economic stripping by conventional surface methods. Quarrying of dimension stone and highwall mining are specialized and less frequently used methods, and will not be looked at in detail in this report.

Aqueous extraction in most cases involves the use of water or a liquid solvent to flush minerals from underground deposits, either by hydraulic disintegration or physico-chemical dissolution. Aqueous extraction includes:

- Placer mining

- Solution mining

Placer mining is intended for the recovery of heavy minerals from alluvial or placer deposits, using water to excavate, transport, and/or concentrate minerals. Solution mining is employed for extracting soluble or fusible minerals using water or a lixiviant.

Open-pit and Cut Mining

Open-pit mining, also known as opencast mining, open-cut mi ing, and strip mining, means a process of digging out rock or mine als from the earth by their elimination from an open pit or borrow.

The word is used to distinguish this type of mining from extractive methods that need tunneling into the earth. Open-pit mines are used when deposits of commercially helpful minerals or rock are found close to the surface; that is, where the overburden (layer material covering the valuable deposit) is comparatively thin or the material of interest is structurally inappropriate for tunneling. For minerals that happen deep underneath the surface-where the overstrain is solid or the mineral happens as veins in hard rock-underground mining methods take out the precious material.

Open-pit mines that manufacture building materials and dimension stone are usually referred to as quarries. People in few of the English-speaking countries are not likely to make a difference among an open-pit mine and other kinds of open-cast mines, like quarries, borrows, placers, and strip mines.

Open-pit mines are characteristically engorged until either the mineral resource is exhausted, or a mounting ratio of overburden to ore makes more mining uneconomic. When this occurs, the exhausted mines are at times converted to landfills for disposal of solid wastes. Nevertheless, some form of water control is normally required to keep the mine pit from becoming a lake.

Open cut mines are dug on benches, which portray vertical levels of the hole. These benches are normally on four meter to sixty meter intervals, relying on the size of the machinery that is being utilized. A lot of quarries do not use benches, as they are normally shallow.

Most walls of the pit are normally dug on an angle less than vertical, to avert and lessen damage and hazard from rock falls. This relies on how weathered the rocks are, and the kind of rock, and also how a lot of structural weaknesses happen within the rocks, like a fault, shears, joints or foliations.

The walls are stepped. The inclined part of the wall is called the batter, and the flat part of the step is called as the bench or perm. The steps in the walls help avert rock falls continuing down the entire face of the wall. In some instances additional ground support is needed and rock bolts, cable bolts and shotcrete are utilized. De-watering bores might be used to ease water pressure by drilling horizontally into the wall, which is frequently sufficient to cause failures in the wall by itself.

A haul road is located at the side of the pit, forming a ramp up which trucks may drive, taking ore and waste rock.

Waste rock is piled up at the surface, near the edge of the open cut. This is known as the waste dump. The waste dump is also tiered and stepped, to lessen degradation.

Ore which has been processed is called as tailings, and is normally slurry. This is pumped to a tailings dam or settling pond, where the water fades away. Tailings dams may frequently be toxic due to the presence of unextracted sulfide minerals, few types of toxic minerals in the gangue, and frequently cyanide which is utilized to treat gold ore via the cyanide leach method.

Pros & Cons of Open Pit Mining

Pros of Open pit Mining

Easier Extraction

The biggest advantage of open pit mining is the relatively low cost-extraction ratio. Open-pit mines are chosen when deposits of valuable minerals are found beneath the surface and where there overburden (overlaying rock or earth) is relatively thin. This means that no extensive tunnel network is required and no costly structural supports. Furthermore, large trucks can enter an open-pit mine, allowing for a more efficient transition from extraction to processing.

Safer Working Conditions

As open-pit mining requires no underground infrastructure, injury rates among workers are significantly lower. Cave-ins are virtually eliminated as risks; and build-up of toxic gases, which can cause sudden explosions or contribute to chronic illnesses, does not occur.

Consumer Benefits

Open-pit mining is significantly less costly than underground mining. Infrastructure and labour savings are passed onto the buyer of mined materials, which eventually trickles down to the consumer.

Cons of Open Pit Mining

Environmental Contamination

As with all forms of large-scale mineral extraction, open-pit mining can have a negative impact on the surrounding environment and ecosystems. The removal of the overburden destroys the pre-existing landscape and contributes to erosion. More-over, chemicals used to treat extracted minerals - for instance, cyanide used to treat gold ore - can sometimes leak into the surrounding soil and water systems. Open-pit mines also require numerous roads and an accompanying production or treatment infrastructure.

Human Health Risks

Tailing ponds are large contained bodies of waste water left behind after mineral ex-traction and treatment. Contaminated water sits in these ponds, which are usually lined with an impermeable material, until the water evaporates and the solid contaminants can be removed. However, these tailing ponds cannot always contain this waste if im-properly made or maintained. This can result in the contaminants leeching into the soil or local surface and groundwater systems. Exposure to many of the chemicals used in mining can cause both immediate and chronic health problems.

Strip Mining

Strip mining, removal of soil and rock (overburden) above a layer or seam (particularly coal), followed by the removal of the exposed mineral.

The common strip-mining techniques are classified as area mining or contour mining on the basis of the deposit geometry and type. The cycle of operations for both techniques consists of vegetation clearing, soil removal, drilling and blasting of overburden (if need-ed), stripping, removal of the coal or other mineral commodity, and reclamation.

Area mining is appropriate for the extraction of near-surface, relatively flat-lying, and thin deposits of coal, phosphate, and similar minerals. Area mining usually progresses in a series of parallel deep trenches referred to as furrows or strips. The length of these strips may be hundreds of metres. Contour mining progresses in a narrow zone follow-ing the outcrop of a mineral seam in mountainous terrain.

In the past, strip-mined mineral deposits that became exhausted or uneconomical to mine often were simply abandoned. The result was a barren sawtooth, lunarlike landscape of spoil piles hostile to natural vegetation and generally unsuitable for any immediate land use. Such spoil areas are now routinely reclaimed and permanent vegetation reestablished as an integral part of surface-mining operations. Generally, reclamation is performed concurrently with mining.

Pros of Strip Mining

- It is very cost effective.

  Mining happens from the first moment work begins with a strip mine. There's no need to develop ground-based infrastructure.

- It offers huge recovery rates.

  Because all of the ground is being directly examined in a strip mine, there is a greater chance of return on an investment. Compared to other forms of mining, a strip mine is up to 40 percentage points more efficient. This helps economies get the resources they need while investors make a profit.

- It allows for products to be brought to market faster.

  Since it is so much easier to recover needed deposits through strip mining, products can hit the marketplace faster. This creates additional economic benefits through spending, creates jobs, and allows for growth.

- Cons of Strip Mining.

  It requires an investment gamble. The processes of strip mining are cheaper. The actual investment into those processes, however, are typically higher than in other forms of mining. From equipment to staffing to distribution, it may not always be possible to gain an exit out of this investment.

- It may contaminate the local water supplies.

  To access some of the needed products in the ground, solvents and other chemicals are sometimes used to gather the needed items. These can then enter into the groundwater, leak into the environment, and pollute the surrounding atmosphere.

- It takes a toll on the environment.

  There's no getting around the fact that a strip mine is essentially just a really big hole. This makes it difficult for the environment to recover from the activities because the resources are being manipulated in an artificial way. Add in flooding, erosion, and other natural events and the area may become unusable after the mine disappears.

Mountaintop Mining

Mountaintop mining is a practice where the tops of mountains are removed, allowing for almost complete recovery of coal seams while reducing the number of workers required to a fraction of what conventional methods require. Mountaintop mining can involve removing 500 feet or more of the summit to get at the buried seams of coal. The earth from the mountaintop is then moved into neighboring valleys.

There are 6 main components of the mountaintop removal process:

Clearing

Before mining can begin, all topsoil and vegetation must be removed. Because coal companies frequently are responding to short-term fluctuations in the price of coal, the trees are often not used commercially, but instead are burned or sometimes illegally dumped into valleys.

Blasting

Many Appalachian coal seams lie deep beneath the surface of the mountains. Accessing these seams can require the removal of 600 feet or more of elevation. Blowing up this much mountain is accomplished by using millions of pounds of explosives.

Digging

Coal and debris are removed using enormous earth-moving machines known as drag-lines, which stand 22 stories high and can hold 24 compact cars their buckets. These machines can cost up to $100 million, but are favored by coal companies because they displace the need for hundreds of miners.

Dumping Waste

In 2002, the Bush Administration changed the definition of "fill material" in the Clean Water Act to include toxic mining waste, which allowed coal companies to legally dump the debris, called "overburden" or "spoil," into nearby valleys. These "valley fills" have buried more than 2,000 miles of headwater streams and polluted many more.

Processing

Coal must be chemically treated before it is shipped to power plants for burning. This processing creates coal slurry, or sludge, a mix of water, coal dust and clay containing toxic heavy metals such as arsenic, mercury, lead and chromium. The coal slurry is often dumped in open impoundments, sometimes built with mining debris, making them very unstable.

Reclamation

While reclamation efforts are required by federal law, coal companies often receive waivers from state agencies with the idea that economic development will occur on the newly flattened land. In reality, most sites receive little more than a spraying of exotic grass seed, and less than 3 percent of reclaimed mountaintop removal sites are used for economic development. According to a U.S. Environmental Protection Agency impact statement on mountaintop removal in Appalachia, it may take hundreds of years for a forest to re-establish itself on the mine site.

Pros of Mountaintop Mining

- It costs less than other methods.

    Mountaintop removal might seem expensive at first glance, considering that it requires large amounts of explosives (which aren't exactly cheap) as well as large, technologically advanced equipment that cost millions of dollars. However, proponents argue that it's actually more cost-effective than traditional mining since it eliminates the need to hire numerous workers. As a result, mining companies can reduce the amount they'd pay on worker salary and benefits (including health insurance, which can be quite expensive since mining is considered to be a high-risk occupation). They can also avoid the bad publicity — as well as the legal costs and hassle — that can arise when their employees get injured or even killed during tunnel cave-ins and other accidents.

- It saves time and effort.

Blasting away several hundred feet of earth is obviously less labor-intensive than traditional mining methods, which means that mining companies can expose coal seams and retrieve coal at a faster and more efficient way. Mountaintop removal also helps mining firms obtain coal from locations that would have been inaccessible if underground mining techniques were used.

- It helps fulfill demands for energy.

With more and more people using electronic gadgets and appliances, the demand for electricity has steadily risen for years. Fortunately, mountaintop removal can help fulfill these demands by helping companies obtain coal (which is widely used in electricity production) in a faster and more efficient way.

Cons of Mountaintop Mining

- It can harm the environment.

Opponents of mountaintop removal argue that it greatly contributes to the degradation of the environment. For one thing, it destroys entire ecosystems since it removes hundreds of acres of forests (which are either cut down or burned), taking away trees and plants that serve as homes and food for many animals. It also promotes pollution since many companies dump topsoil along with toxic mining byproducts and other debris into nearby streams and rivers. On top of that, mountaintop removal introduces air pollution and noise pollution into the area through the large trucks that cart away coal and the explosives that are used to blow up mountains.

- It can harm human health.

Several studies have found that people who live near areas with mountaintop mining develop a wide range of illnesses, including certain kinds of cancer as well as lung, kidney, and heart problems. This isn't really surprising since the blasting process used in mountaintop removal releases dust and toxic chemicals into the air, which can then be inhaled by those who live nearby. The mining byproducts that are deposited in waterways can leach into the area's water table, causing residents to drink polluted water that contains dangerous chemicals.

- It can cost people to lose their jobs.

As mentioned above, one of the main reasons why mountaintop removal is cost-effective is that it requires fewer workers than traditional mining. This can be beneficial for mining companies, but it can be devastating for workers who depend on the mining industry. When companies switch to mountaintop removal, these workers can lose their jobs and will no longer have the means to support themselves and their family.

Dredging

Dredging is a process which is used to remove the deposits percolated underwater for the purpose of clearing the water pathway for ships to pass; to create adequate space to construct important bridges, dykes and dams and to weed out silt, intoxicants and pollutants from the bottom of the water.

Importance of Dredging

Dredging is very important because it helps to moderate the underwater traffic as per the way a dredger wants to moderate it. Additionally, in cases where dredging is specifically used for the purpose of catching shellfish and molluscs, one can find an amazing variety of these creatures with the help of dredging. And, in cases where engineering science needs to be applied, dredging becomes important as without the application of dredging one will not be able to channelize the construction of the requisite edifices properly. Only by making adequate and correct usage of dredging tools and by dredging the right amount of underwater silt and compositions, have a lot of excellent constructions been done so as to be marveled by people across the world.

Uses of Dredging

Sediment removal, or dredging, is used for a number of purposes. Below are the applications that one would use dredging for:

- Waterway Maintenance: Dredging is an important waterway maintenance step, which is probably its most important application. By removing the accumulated debris, dredging can restore the waterway to its original depth and condition. Dredging also removes dead vegetation, pollutants, and trash that have gathered in these areas.

- Create waterways: Many ports are building new waterways with dredging to reach new trade centers and improve the efficiency of the transport of goods. Dredging ensures cargo vessels of all sizes can dock and do not run aground.

- Excavation: Sediment removal plays an important role in the preparation for construction projects such as bridges, docks and piers by performing the necessary underwater excavation work.

- Reclamation: Dredging can remove contaminants that occur due to chemical spills, sewage accumulation, buildup of decayed plant life and storm water runoff.

- Increasing Waterway Depth: As sediment builds up on the bottom of the waterway, it reduces the depth of the water. Dredging strips away the accumulated debris, which can restore the water body to its original depth and reduce the risk of flooding.

- Wildlife Preservation and ecosystem maintenance: Dredging helps ecosystems in a variety of ways. By removing trash, sludge, dead vegetation and other debris, it keeps the water clean and preserves the local wildlife's ecosystems. It also remediates eutrophication, which is an excess of nutrients in the water due to runoff. By solving eutrophication, you stop the excess growth of plant life, which can cause oxygen deprivation.

- Reconfiguring for Larger Ships: By deepening and widening a waterway, dredging can make it passable for larger cargo vessels, which can have a positive economic impact.

- Shore Replenishment: Storms, offshore mining, natural disasters, like hurricanes, and human-made disasters can cause a beachfront to erode over time. Dredging can help to restore the beachfront to its original condition. Beachfront areas often erode which can change their landscape and impact the local ecosystem. Dredging reverses the effects of soil erosion, keeping the local ecosystem and its native plant and aquatic wildlife intact.

- Gathering Construction Materials: The sediment removal process is sometimes used to gather sand, gravel and other debris used to make concrete for construction projects.

- Trash Removal: Dredging can assist in keeping waterways clean by removing trash and debris from beneath the surface.

- Mining for Precious Metals: In certain bodies of water, the sediment can contain traces of precious metals such as gold and diamonds. Dredging can aid in excavating this mineral-filled sediment.

- Pond and Lagoon Cleaning: As ponds and lagoons contain stagnant water, they often can become mucky and have a foul odor. By using dredging, one can remove the accumulated sediment that has caused this making for a healthier body of water.

Working of Dredging

A dredge, the machine used to execute the sediment removal process, is equipped with a powerful submersible pump that relies on suction to excavate the debris. A long tube carries the sediment from the bottom to the surface. The disposal of the dredged material must be conducted in compliance with federal, state and local government laws and regulations.

When dredging, the operator lowers the boom of a dredge to the bottom (or side) of the body of water. A rotating cutter-bar then uses teeth to loosen the settled material, as the submersible pump removes the sediment from the bottom of the waterway. The silt and debris are then transported away for final processing.

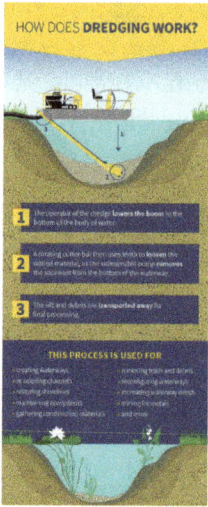

Different Types of Dredges

There are several types of dredges used in the sediment removal process. The most common types of dredges are:

- Plain-Suction: A plain-suction dredge is the most common type of sediment removal equipment. Unlike other dredge versions, it doesn't contain a tool for penetrating or cutting into the bottom of the water body — it relies on suction to remove loose debris.

- Cutter-Suction: This type of dredge contains a cutting tool that loosens material from the bottom and transports it to the mouth of the suction apparatus. The use of a cutter-suction dredge may be necessary for removing debris from hard surfaces that would prevent efficient suction via standard methods.

- Auger-Suction: An auger-suction dredge essentially bores holes into the bed to loosen and suck up the debris. The rotating auger can burrow deeply into the surface. This type of dredge works well for sludge removal applications at wastewater treatment plants and other areas requiring heavy-duty sediment removal.

- Jet-Lift: This technologically advanced sediment removal equipment works by injecting a high-volume stream of water to pull in nearby water, silt, and debris.

Benefits of Dredging

Dredging provides numerous benefits for shipping, construction, and other projects. The advantages of dredging are:

- Widening and Deepening: Dredging can be a critical process for the commercial shipping industry. Removing sediment can maintain the appropriate width and depth for enabling the safe, unobstructed passage of cargo vessels carrying oil, raw materials, and other essential commodities.

- Waterway Project Preparation: Dredging is a critical underwater excavation step in many waterway construction projects such as bridges, docks, piers and underwater tunnels.

- Land Reclamation Projects: Sediment removal is sometimes used as a source of materials for land-building projects. The liberated sediment can then be dried out and transported to a new location where additional land is required for building and other purposes.

Dredging also has numerous environmental benefits, including:

- Environmental Remediation: Sediment removal can help to restore a shoreline or beachfront to its original condition by reversing the effects of soil erosion.

- Cleanup Applications: Dredging can clean up a waterway after a toxic material spill or via the removal of trash, debris, decaying vegetation, sludge or other materials that can contaminate water and soil.

- Preserving Aquatic Life: Dredging can produce a healthier aquatic eco-system that can result in a more suitable habitat for fish and other wildlife. It can also be used for trash and debris removal to keep the waterways clean.

- General Pollutant Removal: Water bodies located near urban areas and industrial complexes can quickly become a receptacle for various pollutants. Sediment removal can prevent the accumulation of pollutants and keep the waterway clean and healthy.

- Remediation of Eutrophied Water Bodies: Eutrophication is an excessive amount of nutrients in a water body typically caused by water runoff from the surrounding land. Eutrophication can lead to an overabundance of plant growth that results in oxygen deprivation and can cause the death of aquatic wildlife. In some cases, dredging may be the most viable remediation option when eutrophication occurs.

Underground Mining

Underground mining is a technique used to access ores and valuable minerals in the ground by digging into the ground to extract them. This is in contrast with techniques like open pit mining, in which the surface layers of ground are scooped away to access deposits, or mountaintop removal, in which the top of a mountain is simply shaved off to access the ore inside. When people think of mining, they often visualize underground mining, and many people think of coal mines in particular, although numerous products can be mined underground, with some of the deepest mines in the world being used to access deposits of gold.

Companies opt to use underground mining when deposits of ore are so buried that surface mining is not an option. This requires a lot of logistics to be safe, even with the highly mechanized nature of modern underground mining. Especially with soft ores, much of the work is done by machine, not people, but it's still important to avoid subsidence above the mine, collapses in the mine, explosions, and an assortment of other health and safety risks.

The first step in underground mining is development mining, in which shafts are dug into the site to make the ore accessible. During this phase, along with shafts, things like electricity are installed, along with lifts and shoring to support the walls of the mine so that it will not collapse. Once the mine is developed, active production mining can begin, with the ore being extracted by hand, by machine, or with a mixture of the two.

In shaft mining, a shaft is dug straight down into the Earth to access buried deposits. People get in and out of the mine through lifts in the shaft, and additional shafts may be dug to create emergency exits and deal with ventilation needs. In slope mining, a sloping shaft is dug so that people access the mine by moving down an incline. Motorized equipment is usually used so that people can get into and out of the mine quickly.

One major need with underground mining is ventilation. Ventilation is necessary to remove dust, byproducts of combustion, and byproducts of explosions, along with gases which may be trapped in pockets inside the Earth. Proper support to prevent collapses is also rather important, as are redundant safety systems including alarms for workers, evacuation shafts, emergency lighting, and so forth.

Two prominent ways through which underground mining is done are: Hard-Rock Underground Mining and Soft-Rock Underground Mining

Hard-Rock Underground Mining

Underground hard rock mining refers to various underground mining techniques used to excavate hard minerals such as those containing metals like gold, copper, zinc, nickel and lead or gems such as diamonds. In contrast soft rock mining refers to excavation of softer minerals such as coal, or oil sands.

Mine Access

Underground Access

Accessing underground ore can be achieved via a decline (ramp), vertical shaft or adit. Declines can be a spiral tunnel which circles either the flank of the deposit or circles around the deposit. The decline begins with a box cut, which is the portal to the surface. Depending on the amount of overburden and quality of bedrock, a galvanized steel culvert may be required for safety purposes.

Shafts are vertical excavations sunk adjacent to an ore body. Shafts are sunk for ore bodies where haulage to surface via truck is not economical. Shaft haulage is more economical than truck haulage at depth, and a mine may have both a decline and a ramp. Adits are horizontal excavations into the side of a hill or mountain. They are used for horizontal or near-horizontal ore bodies where there is no need for a ramp or shaft. Declines are often started from the side of the high wall of an open cut mine when the ore body is of a payable grade sufficient to support an underground mining operation but the strip ratio has become too great to support open cast extraction methods.

Ore Access

Levels are excavated horizontally off the decline or shaft to access the ore body. Stopes are then excavated perpendicular (or near perpendicular) to the level into the ore.

Ventilation

Door for directing ventilation in an old lead mine. The ore hopper at the front is not part of the ventilation.One of the most important aspects of underground hard rock mining is ventilation. Ventilation is required to clear toxic fumes from blasting and removing exhaust fumes from diesel equipment. In deep hot mines ventilation is also required for cooling the workplace for miners. Ventilation raises are excavated to provide ventilation for the workplaces, and can be modified to be used as escape routes in case of emergency.The main sources of heat in underground hard rock mines are virgin rock temperature, machinery, auto compression, and fissure water although other small factors contribute like people breathing, inefficiency of machinery, and blasting operations.

Soft-rock Underground Mining

Underground soft rock mining is a group of underground mining techniques used to extract coal, oil shale, potash and other minerals or geological materials from sedimentary rocks. Because deposits in sedimentary rocks are commonly layered and relatively less hard, the mining methods used differ from those used to mine deposits in igneous or metamorphic rocks.

This method is implemented after a careful study and the amount of deposits available in the area. The economic liability of the project is also taken into consideration before going ahead as the investment into the scheme will be large and returns should commensurate with the amount of spend on the project.

Underground soft mining involves blast mining, which is an essential part of mining. In almost all forms of mining we can see that blasting and drilling occurs. It is a method to break the material into smaller parts by blasting technology. Blast design and the execution of this process are equally important in this method. If this method is not executed in the proper way the economics of the mine will be drastically affected. This method is also used in open pit mining method. The blasting of a mine is done to break a large quantity of the block to disintegrate and later remove such blocks by using equipments to carry the minerals to the dump site.

The underground soft mining also made use of short wall method. This method was developed in the 1960's and had lot of advantages. The main advantage of this method was that it can cut the cost of mining by not installing completely the long wall devices. In some cases more flexible short wall mining was done to obtain the benefits of long wall mining. However this method was not very successful in the long run. The short wall method was used to extract the minerals in a comparatively less space but the cost effectiveness is almost the same. The coal skimming was another method introduced to extract coal in a more simple way. This method was much faster and easy when compared to the previous methods. Level surface drilling jumbo called single boom underground drilling jumbo works mainly on the surface of the tunnel. It has the capacity to drill both the bolt hole and the blast hole.

When it comes to mining, the most important factor that has to be taken into consideration is the equipment used for mining. Be it underground or surface mining, the equipment deployed for various types mining differs on the minerals to be extracted and it plays a major role in the successful implementation of the project. By making use of these methods the mining industry is contributing a lot to the economic development of the country.

Underground Mining Methods

Longwall Mining

In the method of secondary extraction known as longwall mining a relatively long mining face (typically in the range 100 to 300m but may be longer) is created by driving a roadway at right angles between two roadways that form the sides of the longwall block, with one rib of this new roadway forming the longwall face. Once the longwall face equipment has been installed, coal can be extracted along the full length of the face in slices of a given width (referred to as a "web" of coal). The modern longwall face is supported by hydraulically powered supports and these supports are progressively moved across to support the newly extracted face as slices are taken, allowing the section where the coal had previously been excavated and supported to collapse (becoming a goaf).

A coal haulage system is installed across the face, on modern faces an "armoured face conveyor or AFC". The roadways which form the sides of the block are referred to as "gate roads". The roadway in which the main panel conveyor is installed is referred to as the "main gate" (or "maingate"), with the roadway at the opposite end being referred to as the "tail gate" (or "tailgate") roadway.

The benefits of longwall mining compared to other methods of pillar extraction are:

- Permanent supports are only needed in the first workings portion and during installation and recovery operations. Other roof supports (longwall chocks or shields on modern longwalls) are moved and relocated with the face equipment.

- Resource recovery is very high - in theory 100% of the block of coal being extracted, though in practice there is always some coal spillage or leakage off the face haulage system lost into the goaf, especially if there is a lot of water on the face

- Longwall mining systems are capable of producing significant outputs from a single longwall face – 8 million tones per annum or more.

- When operating correctly the coal is mined in a systematic, relatively continuous and repetitive process which is ideal for strata control and for associated mining operations

- Labour costs/tonne produced are relatively low

Disadvantages are:

- There is a high capital cost for equipment, though probably not as high as first appears when compared to the number of continuous miner units which would be required to produce the same output.

- Operations are Very Concentrated ("all eggs in one Basket").

- Longwalls are not very flexible and are "unforgiving" - they do not handle seam discontinuities well; gate roads have to be driven to high standards or problems will arise; good face conditions often depend on production being more or less continuous, so problems which cause delays can compound into major events.

- Because of the unforgiving nature of longwalls, experienced labour is essential for successful operations.

Shortwall Mining

This method of mining was developed in the late 1960's to take advantage of the then recent development of suitable hydraulic longwall supports, coupled with the productivity and low capital cost of continuous miners and shuttle cars. In effect it gained some of the advantages of longwall mining without the cost of installing a complete set of longwall equipment.

An installation roadway was driven as for a normal longwall, but only supports were installed. A continuous miner was then utilised to cut 3.5m wide open ended lifts off the face, with shuttle cars being used to transport coal along and off the face to the maingate belt in lieu of an AFC.

The face length was therefore limited by the length of shuttle car cables then available, but in practice most shortwall faces were considerably shorter than this (<90m).

Supports were connected to a reference rail which was then utilized to pull the 2 or 3 leg supports forward, in a similar manner to the use of an AFC to advance longwall supports.

Shortwall faces could be installed between two gate roads as for a longwall face, but in some cases were mined to a blind end and ventilated by auxiliary fan (not very suitable for gassy seams as the fans could draw from the goaf).

Shortwalls were used in an endeavour to increase the productivity of continuous miners at relatively low capital cost, sometimes as a transition stage while changing a mine to full longwall. In some cases, because they were somewhat more flexible, shortwalls were used to obtain the benefits of longwalls in mines, or parts of mines, where seam discontinuities or mine geometry made the use of full longwalls impractical.

The main disadvantages of shortwalls compared to longwalls are:

- The width of the unsupported roof ahead of the chocks is governed by the width of a continuous miner as opposed to a shearer drum.

- Personnel have to work adjacent to the face which presents safety issues unless rib support is installed which would greatly slow production.

- The use of shuttle cars is by its nature not continuous and brings in all the disadvantages of trailing cables in the face area.

Shortwalls had only mixed success and there have been no shortwall operations in Australia (or elsewhere to the knowledge of the writer) for many years.

Cut and Fill Mining

Cut and fill mining is a highly selective open-stope mining method considered ideal for steeply dipping high grade deposits found in weak host rock. Many variations of the general cut and fill technique exist, however this article will focus on overhand cut and fill. Overhand cut and fill evolved from square set stull stoping to provide stronger support. In this method, mining begins at the bottom of the ore body or block and progresses upward. During the mining sequence, the back of the excavation is temporarily supported using rock bolts before the stope is back filled to form the floor of the next level of development. Backfill is designed to provide mild excavation support as well as to provide a strong working floor for personnel and equipment. Backfill selection is dependent on the quality of the host rock and the size of equipment operating a top of the backfill. Progression between stopes is achieved through the construction of raises driven upward through the ore body.

Selection Considerations

There are a number of ore body features that should be considered when evaluating the applicability of a cut and fill mining technique. These features include:

- Orebody geometry
- Ore grade/method cost
- Rock quality
- Waste Disposal

The ore body must be narrow and steeply dipping as the method relies on gravity to draw ore. Cut and fill is a very expensive mining method, due in large part to the high costs of backfill. This cost can be justified if the ore body is of a particularly high grade. (Cut and fill, Drift and fill) The rock quality designation is a further critical factor. Though the surrounding country rock may be weak, and require support following excavation, the ore should exhibit strength qualities that make it safe enough to be worked under or allow it to be supported sufficiently. In general, cut and fill can be an attractive option surface tailings disposal must be minimized since much of the waste rock can be returned underground for use as backfill.

Planning

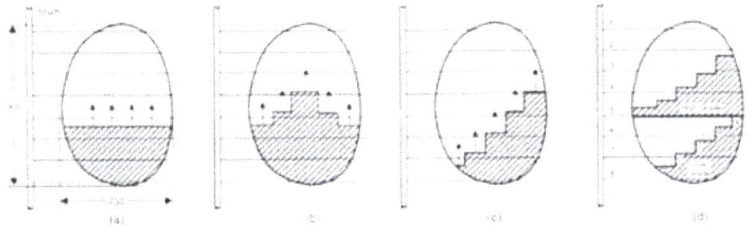

The first step in the planning process for a cut and fill operation is to determine the type of backfill that will be used. Past fill and hydraulic fill draw additional consideration in terms of planning due to the infrastructure requirements such as paste plants, pumping systems, and piping networks. The chemical characterisitcs of the waste rock must also be considered since these effect its efficacy as a filler element. Floor dilution generated from excessively uncosolidated or weak fill must also be considered. Secondly, access selection (ramp or shaft) may be of importance in the planning process since a ramp may offer the flexibility of continuous mining by reducing the cyclical effect of mining. Thirdly, a preliminary selection of material removal equipment can be made. The geometry of the ore body dictates the size and type of such equipment including slushers, drills, and LHD's. Cut and fill operations with large stope geometry may permit the use of drill jumbos and large LDH equipment whereas smaller stopes may preclude the use of stopper drills and smaller LHD's. This selection is also dependent on the desired productionr ate and ore-pass capacity. In cut and fill mining all work is done inside the stope therefore proper ventilation must be provided to the worker. To accomplish this, adequately sized ventilation raises must be constructed. Given the selective nature of cut and fill, the stope geometry is largely dictated by the ore body shape, thus the design of each stope will be unique and may be subject to great variation even within a designated zone.

Ground Control and Rock Mechanic Consideration in Stope Planning

Given the high degree of worker expsoure in cut and fill operations, rock mechanics properties must be extensively known in order to provide adequate protection to workers. In the case of deep mining, where stresses are high, the most economical strategy for ground control is proper stope planning. A staggered stope face advance is used to reduce the stress envelope experience at the mining face. The figure below demonstrates this staggering technique used to reduce stress concentrations. In addition, temporary support may be required to secure rock along near the excavation surface while workers are present.

Development

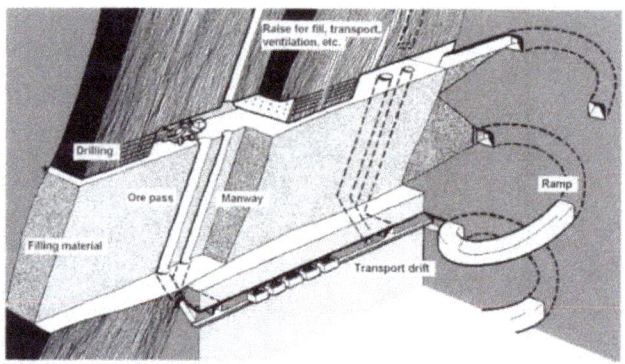

Cut and Fill Development

The development sequence described below pertains to a highly mechanized over-head cut-and-fill operation where the stope is mined from the bottom upward. This technique is illustrated in the figure:

- An undercut is constructed beneath the stope along its entire strike length. This undercut will form the transport drift from which ore will be removed by LHD's or rail cars and will also provide access for construction of draw points.

- A ramp is constructed at the side of the stope connecting the transport drift and the first production level. This ramp will provide access for drill jumbos and LHD's to the production level.

- A cut is made beginning at the intersection of the ramp and stope to form the first production level. The width of the pillar between the roof of the transport drift and the floor of the production levels is highly dependent on rock mechanics characteristics of the ore.

- An ore pass is constructed through the floor pillar of the production level to connect the transport drift to the production level.

- A manway is constructed at a location near to the ore pass to connect the production level to connect the transport drift to the production level.

- Additional ore passes and manways are constructed at regular intervals along the strike length of the stope as the production level advances horizontally.

- Auxiliary ventilation fans are installed in the access ways as required delivering fresh air to the working face. These fans may be in the range of 5-15HP depending on the size of the face and the number and size of equipment in the stope.

- Using a raise boring machine, a raise is constructed connecting the production level to a point where backfill may be fed; either an upper level or surface. This raise may also act as a ventilation raise.

- The drill pattern is dependent on the drilling equipment employed. A typical jumbo may drill holes 3m deep spaced in a 1m x 1m grid. The drill pattern can be tailored to meet the desired production rate by adjusting the volume of broken rock produced by each blast sequence. The figure shows a cross-section of a typical drill pattern.

- Using LHD equipment, the broken rock is dumped down the ore-pass travelling down to the transportation drift where it is gathered at a collection point.

- The drill, blast, and muck sequence is repeated until mining has progressed along the entire strike of the stope.

- Cribbing is placed at the top of the existing manways and ore passes to form a lining in preparation for backfilling. Old ventilation tubing may be used for this purpose.

- Backfill enters the stope through the previously constructed raise in the production level roof. The level is filled and allowed to dry. Drying time is dependent on the backfill type and moisture content.

- The ramp at the side of the stope is extended to the next production level to provide entry for drill equipment. The cycle is repeated until the vertical extent of the stope is reached.

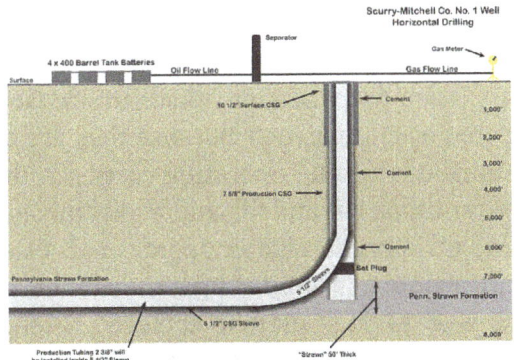

Drill Patern for Cut and Fill

Backfill Requirements

There are a number of options available for the backfill to be used in cut and fill mining, the choice of which is dependent on the support requirements of the area. These options include:

- waste fill

- pneumatic fill

- hydraulic fill with dilute slurry

- high-density hydraulic fill (paste fill)

The highest strength option available is paste fill, followed by sand fill, and finally unconsolidated rock fill. Beyond the support requirements, the fill must be able to support any equipment that is necessary for stope development, as it will become the working floor for the next stope.

Costing Information

When calculating the cost of a cut and fill operation, both primary and secondary development must be considered. Primary development work includes development of the main access shaft or ramp, secondary/escape shaft or ramp, level development, pump rooms, and hoist rooms. The secondary development costs consists of developing the sublevels, footwall ramps, raise preparations, raises to ore, access crosscuts, and ore passes; these are considered operating costs because they are ongoing. Once developments have been accomplished, mining operating costs can differ considerably from overhand cut and fill, conventional cut and fill, and highly mechanized cut and fill. For a highly mechanized cut and fill operation major operational costs include drilling, blasting, raising, mucking and slushing, ground support, cleanout, preparation for backfill, and backfilling material. Cost savings may be achieved through increased mechanization. Subsequently, increased mechanization provides a safer work environment as worker exposure to dangers at the working face is limited.

Table: Comparison of labor requirements for hourly personnel

Job Classification	Conventional* Method	Highly Mechanized* Method
Electrician	4	4
Welder	3	3
Warehouse worker	3	2
Lamp operator	2	2
Janitor	1	1
Laborer	4	4
Total Surface	17	16
Surface shaft worker	4	4
Hoister	4	4
Underground shaft Worker	6	6
Grizzly Tender	2	2
Motor operator	8	8
Track maintenance	3	3
Utility workers	4	4
Laborers	4	4
Total underground service	35	35
Haulage drifts	8	8
Raises	8	2
Total development miners	16	10
Stope miners	64	18
Fill crew	4	6
Total miners	68	24
Total hourly	136	85
Total cost at $10.50 per hr□ x 1920 hr per year	$2,741,760	$1,713,600

* Based on 2 shifts per day, 2 operating levels, 8 stopes per level, and 2 miners per stope crew.

† Based on 2 shifts per day, 3 stoping areas, and 3 miners per stoping areas.

‡ Includes burden.

Costing Information for Cut and Fill

Table: Comparison of labor requirements for total personnel

	Conventional Method		Highly Mechanized Methods	
	No.	Annual Cost. $	No.	Annual Cost. $
Salaried	29	716,850	22	556,200
Hourly	136	2,741,760	85	1,713,600
Total	165	3,458,610	107	2,269,800
Miner-shifts per day		39,600		25,680
Annual production		192,000 st*		192,000 st
Labor cost per short ton		18.01		11.82
Short tons per miner shift		4.85		7.48

* Metric equivalent: st x 0.9071847 = t

Costing Information for Cut and Fill

Infrastructure Requirements

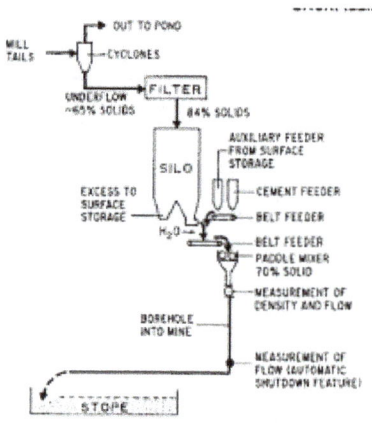

Paste Plant Flow Sheet

Cut and fill mining requires a number of unique infrastructural requirements mostly pertaining to bakcfill preparation and delivery. These requirements are dependent on which back fill method is chosen. For paste or hydraulic backfill, the mine site must include a backfill plant and the underground network used to deliver the backfill to the working stopes. The underground network will include piping down to each level, and moveable pipes on each level to reach the individual stopes. Many mines use gravity to drive the delivery system, however sometimes it is necessary to introduce pumps into the system. The nature of the backfill means that wear will occur on the pipes, and their condition must be monitored in order to ensure that the system does not encounter down time when it is most needed. An example of the backfill infrastructure required for paste or hydraulic backfill is displayed below. The infrastructure required for rock fill is somewhat different, typically requiring mechanical dumping access so that LHDs or trucks can transport and deliver crushed rock fill.

Advantages and Disadvantages

Advantages

- High selectivity and low dilution ma achieved.
- Minimal development is required; low capital cost.
- Versatile for mining method; can follow irregular orebodies.
- Flexible; mining method can be easily modified.
- Low equipment investment relative to other methods.
- Minimizes ground movement.

Disadvantages

- Cyclical ore production.
- Labour and skill intensive.
- Dangerous working conditions; work conucted a top freshly blasted rock.
- High degree of ground control required.
- Expensive and costly ventilation system.
- Need for backfill infrastructure (piping and paste plant).
- Not suitable for low grade ore due to high mining cost.

Room and Pillar Mining

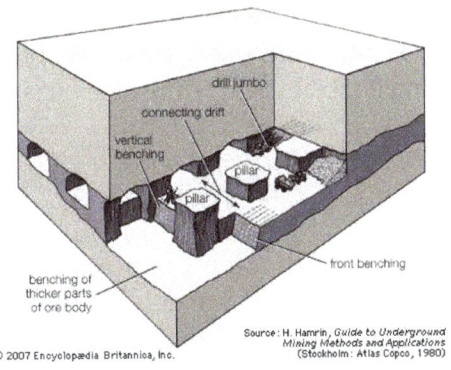

Room and Pillar Mining

Room and pillar is an underground mining method that has applications to a wide variety of hard-rock deposits worldwide. It is commonly classified as an open-stoping method, meaning that development involves mining out underground cavities while leaving the surrounding un-mined waste or ore as primary support. Room and pillar is differentiated from other open-stoping methods, in that the support rock typically extends from hangingwall to footwall in the form of pillars. Pillars are usually round or rectangular and are completely surrounded by open excavations called 'rooms'. The mining method often warrants the use of secondary support such as rockbolts, reinforcement rods, and shotcrete but this does not preclude its classification as an open-stoping method. There are four main variations of room and pillar: classic, post, step, and steep room and pillar.

Room and pillar is one of the oldest existing mining methods, dating back over 1000 years. In its early use, detailed stope planning was very uncommon and mine operators would typically follow apparent high grade areas, leaving pillars only when necessary to stabilize openings. In early 20th century America, when this lack of planning still existed, room and pillar mining was referred to as 'gophering' because of the random and asymmetrical pillars that resulted from development. Today's mining industry is more systematic and most current room and pillar mines go through rigorous planning prior to development. As will be discussed more thoroughly in the planning section, early development can have a profound effect on the late stages of mining and project economics; especially in multipass mining systems.

Room and pillar has a rich history in soft-rock mining and is commonly associated with coal, potash, uranium, and other industrial materials. Although, it is not uncommon to see room and pillar used in hard-rock metal mines for such commodities as lead, zinc, and copper. A few noteworthy areas in North America that have utilized hard-rock room and pillar mining are Missouri, Tennessee, Nanisivik(Baffin Island), Sudbury (Ontario), and Elliott Lake(Ontario). The latter three of the aforementioned areas, all of which are located in Canada, employed post-pillar mining.

Deposits that are exploited with room and pillar are usually sedimentary-bedded deposits because of their characteristic properties of being flat with large horizontal extent are ideal for room and pillar mining. Although, room and pillar is very versatile and can be applied to non-conforming deposits using different variations or in conjunction with other mining methods.

Geological Requirements

Ideal conditions for room and pillar mining are regular, flat orebodies with large horizontal extent and competent ore and waste rock, however these are by no means limiting criteria. Room and pillar is considered one of the most versatile mining methods. To illustrate its versatility, room and pillar has been used in deposits with rock strengths ranging from 30 to 350 MPa, depths ranging from 15 to 900m, and inclines ranging from 0 to 55 degrees. That being the said, the profitability of room and pillar is very sensitive to the aforementioned parameters and it is important to understand the size, shape, incline, and grade continuity of a deposit before determining its suitability for room and pillar mining.

Strength

There are no strict guidelines for minimum ore and waste rock strength but ultimately, for room and pillar to be viable, the orebody should be competent enough to support itself without significant ground support. Therefore, any deposit that could support an open excavation without major ground support could be feasibly mined with some form of room and pillar. However, there inevitably becomes a point in some deposits where the necessary pillar size to support the mine becomes so large that room and pillar stops being the best economic option.

Pillar size and shape are designed based off strength and stress estimates from geotechnical information and are designed to be as small as possible to maximize recovery while ensuring a safe working environment. Some pillars must be able to provide support long enough to complete stope extraction while other pillars, located in more critical areas, must be designed to provide support for the duration of the mine life. In appropriate rock conditions, pillars can be designed to fail gradually under close monitoring but this is very dependent on stress conditions. At very large depths below surface, in hard-rock conditions, some pillars are able to absorb massive amounts of energy before deformation and fail violently.

Geometry

The geometry (size, shape, thickness) of the deposit are crucial in determining the suitability of room and pillar. Room and pillar can be feasibility utilized on a number of different types of deposits, however it is usually is not recommended for use in steep deposits (>55 degrees), where material is able to flow by gravity. This is especially the case in very thin (rooms become too small for mechanization) or very thick deposits (pillars become too high to support the open stopes). There are some case when relatively steeply mined orebodies with large vertical extent can be mined (post-pillar) but orebody inclination normally does not exceed 55 degrees. Room and pillar also becomes difficult in deposits with large vertical extent. Extremely large pillars cause problems with deterioration, and the roof becomes hard to monitor and maintain.

Most hard-rock room and pillar mines are large (2,000 to 7,000 tpd) but there are some very small zinc mines in Illinois-Wisconsin that use room and pillar. One common consistency between these small and large deposits are regular dimensions. When thickness is constantly changing, it is difficult to use room and pillar. Pillars of different sizes pose rock mechanics issues and make stress conditions very difficult to simulate during planning. Although, there are some obscure cases when room and pillar has been used in orebodies with both small and large veins. In Tennessee, there are some mines which follow very narrow veins and then open up into large collapsed dome structures, where stoping must expand both horizontally and vertically to fully exploit the areas.

Continuity

Continuity is also an important geological aspect for room and pillar evaluation. Pillars are usually designed in regular patterns with consistent dimensions to one, make development manageable, two, make simulation of stress conditions easy, and three, keep haulage roads straight to make mechanization efficient. If the ore grade is continuous, regular pillar layouts are easy to justify but complications arise with inconsistencies in ore grade. When it is difficult to predict the grade throughout the deposit, it is difficult to produce an economically optimal pillar layout.

Geotechnical

To design pillars to a size where they are able to stabilize an opening without excessive ground support, a detailed geotechnical study should be carried out, documenting roof and ore strength, geological structures, joints and fractures (rock mass classification), and in-situ stress measurements. With these parameters, rock-mechanics engineers can anticipate pillar behaviour in different configurations and design them accordingly. The following table shows common rules of thumb for safety factor in pillar design.

Safety Factor	Pillar type
2	Pillar located in development headings.
1.1-1.3	Panel pillars after retreat mining.
1.0	Pillars which are planned to fail.

Pillar design factors

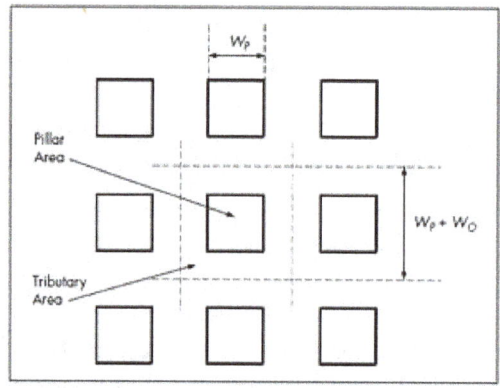

Tributary loading diagram

The stress on a particular pillar can be calculated using the tributary load equations.

Tributary loading is the phenomenon where stress accumulates towards structural members (pillars). The image shows a diagram, explaining tributary loading area.

With a regular pillar layout, pillar stress can be calculated in following way:

$$\sigma = \gamma \times z$$

$$\sigma_p = \sigma \times \frac{1}{1-R} \; with \; R = \frac{A_t - A_p}{A_t}$$

Where:

σ refers to overburden stress

γ refers to specific gravity

z refers to depth below surface

σ_p refers to pillar stress

A_t refers to tributary area

A_p refers to pillar area

Pillar Orientation

From the geotechnical study, it is also important to determine the orientation of in-situ stresses. Pillars should be orientated in the direction of the maximum stress. This is especially of concern in deep hard-rock mines with tremendous horizontal stresses.

Pillar Rating	Pillar Condition	Appearance
1	No indication of stress-induced fracturing. Intact pillars.	
2	Spalling on pillar corners; minor spalling of pillar walls. Fractures oriented subparallel in walls and are short relative to pillar height.	
3	Increased corner spalling. Fractures on pillar walls more numerous and continuous. Fractures oriented subparallel to pillar walls and lengths are less than pillar height.	
4	Continuous, subparallel, open fractures along pillar walls. Early development of diagonal fractures (start of hourglassing). Fracture lengths are greater than half of pillar height.	
5	Continuous, subparallel, open fractures along pillar walls. Well-developed diagonal fractures (classic hourglassing). Fracture lengths are greater than half the pillar height.	
6	Failed pillar, may have minimal residual load-carrying capacity and be providing local support to the stope back. Extreme hourglassed shape or major blocks have fallen out.	

Vibernum Trend pillar classification system.

Pillar Rating System

To evaluate the condition of existing pillars, mines often develop a pillar rating system in hard rock mines. As an example, the mines of Vibernum Trend use a rating system from 1 to 6 to monitor pillar condition. A rating of 1, indicates no signs of stress and a rating of 6, indicates complete failure. The figure shows images representing the different pillar conditions. The maintenance policy at Vibernum Trend requires pillars reaching 3 on the rating scale to be reinforced to prevent further deterioration.

An important part of pillar maintenance involves monitoring convergence. Convergence is good predictor of pillar failure and can help engineers respond before convergence accelerates uncontrollably.

Planning Considerations

The feasibility of room and pillar is subject to a detailed planning process that must aim to maximize net present value (NPV) while maintaining a safe working environment. The process does not have a unique methodology, varying quite significantly based on orebody characteristics. The fundamental key to creating a room and pillar mine plan is understanding the relationship between all the different economic factors. The most important factors being safety, recovery, ground support, efficiency, and legal requirements.

Pillar Size

One of the steps of the planning process is understanding the trade-offs between recovery and ground support and ultimately finding a pillar size that is economically optimal. As pillar size decreases, reserves increase but so does the need for ground support. The point at which the added value of decreasing the size of pillars equals the cost increase for the necessary additional ground support, is the point where optimum pillar size has been reached. In reality, orebodies are not homogeneous throughout, therefore this analysis must be done on each section of the mine and in some cases, each individual pillar.

In the pillar size optimization there are a few other dependent factors that must be considered. Two of these aspects are ventilation and equipment. As pillar size decreases, room size increases which changes ventilation requirements and equipment options. Larger rooms oblige bigger fans to achieve minimum air flow velocity but allow for much larger equipment which can improve efficiency and operating costs. However, these factors are not just dependent on pillar size but also on pillar layout.

Pillar Layout

Regular pillar layouts improve mechanization efficiency and decrease ventilation requirements, however depending on the distribution of economic and deleterious materials, irregular layouts can sometimes be justified. Regular pillar layouts create more efficient roads for mechanized equipment to travel on. Optimal mechanization planning works to minimize haulage grades, and keep haul roads straight as possible and in excellent conditions, while avoiding the need for abrupt turns. Regular pillar layouts also create less resistance to air flow which decreases flow requirements. There is also the added complication of pillar recovery in this analysis.

In the resulting plan, all federal and local legal requirements must be met. This may include a minimum pillar safety factor, minimum flow/velocity air requirements, and maximum pillar height. Government authorities should be contacted prior to development.

Room and pillar planning is usually fairly straightforward in deposits where grade and thickness is consistent throughout, as is typical of coal and potash deposits. Planning, however, becomes considerably more complicated in hard-rock metal mines where there can be considerable deviation in the shape, grade, and strength of the orebody. These plans must then be flexible to change but this is often difficult in open-stoping mines where pillars are dependent on one another for support.

Multipass Mining

In room and pillar mining, engineers have a choice of whether to take the whole orebody in one slice or in multiple slices. The need for multiple slices can arise when the orebody is very thick and the pillars cannot support the full height of the deposit. Multipass mining is also used in mines where there is uncertainty in stress conditions and the engineer decides to take a more cautious first-pass and determine from there what the best course of action would be. In hard-rock mines, it is often to difficult to verify the exact thickness of a deposit because of poor continuity. When this is the case, it is difficult to decide between single and multi-pass room and pillar and it is typically recommended to begin on the projected top most slice to make it easy for the back to be reached.

Variations and Development

Development for room and pillar mining methods is generally quite small. The bulk of the development takes place within the ore body, therefore a mine is producing ore as development is carried out. Depending on the geometry of the ore body, haulage drifts and access ramps may need to be developed in the footwall and truck loading bays may need to be developed into waste rock if the thickness of the ore does not allow for the necessary height to load trucks.

There are four main variations of room and pillar mining which are differentiated by their applicability to different shaped orebodies. Classic room and pillar is applied to flat-bedded deposits, post-pillar is used for inclined orebodies with large vertical extent, and step and steep room and pillar are used in inclined orebodies with limited thickness. A useful video showing classic room and pillar development can be seen here.

Classic

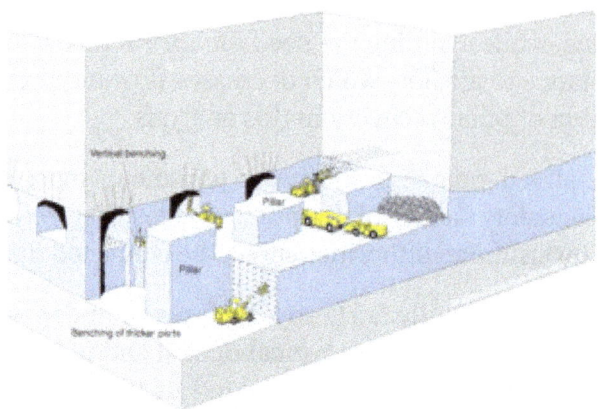

Development in classic room and pillar.

This is the most generic room and pillar mining, applied to flat regular orebodies, usually with substantial thickness. Development consists of making large open stopes with trackless equipment operating on the floor. This method usually has high utilization and efficiency and planning is usually relatively simple.

When starting ore production, it is normal practice to drive a drift in the ore that will allow the mine to open about four or five rooms off of the initial drift. Since this initial drift will serve as the main haulage way, it is important to be cognizant of road grade and make an effort to keep them as flat and straight as possible. The next step in production is to mark off the pillars that will be left in place, as identified by the mine planner. At this point, the rooms can be driven off laterally from the main drift. If the mineralized region extends beyond the length of the main haulage drift, care should be taken to ensure that the drift continues straight and pillars are not intersecting at any point. In many cases, "doglegs" will occur where the haulage drift is forced to weave around these pillars which increases transportation times and results in more expensive haulage. The image shows typical classic room and pillar development.

When starting ore production, it is normal practice to drive a drift in the ore that will allow the mine to open about four or five rooms off of the initial drift. Since this initial drift will serve as the main haulage way, it is important to be cognizant of road grade and make an effort to keep them as flat and straight as possible. The next step in production is to mark off the pillars that will be left in place, as identified by the mine planner. At this point, the rooms can be driven off laterally from the main drift. If the mineralized region extends beyond the length of the main haulage drift, care should be taken to ensure that the drift continues straight and pillars are not intersecting at any point. In many cases, "doglegs" will occur where the haulage drift is forced to weave around these pillars which increases transportation times and results in more expensive haulage.

Post

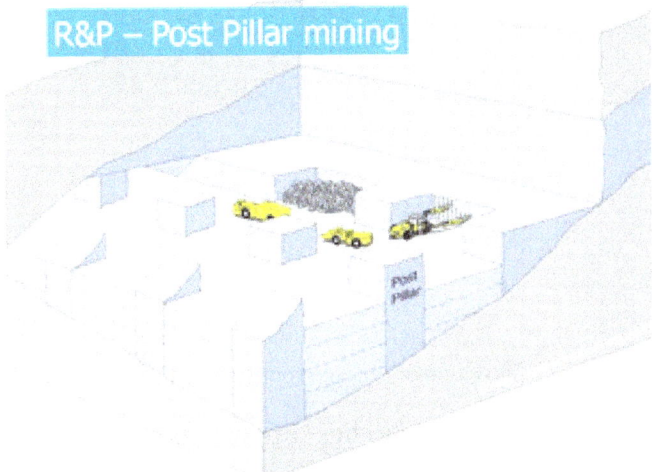

Development in post pillar mining.

Post pillar is a lesser known room and pillar variation that is applied to orebodies dipping from 20-55 degrees with large vertical extent. The ore is mined in a series of horizontal slices. Mining development progresses upward from the bottom horizon and the rooms are backfilled with backfill used on the next pass as the working floor (essentially a modified cut and fill). The pillars are left in the same location at each subsequent horizontal level as production moves upward. It is possible for production to move downwards but this is very uncommon as the backfill would need to be very strong, with the backfill becoming the back rather than the floor of the next subsequent horizontal stope.

The development into the ore body for the post-pillar mining method is similar to classic room and pillar on the first pass. A main haulage drift is driven into the ore body which will provide enough working faces for mining operations to be carried out efficiently. The main difference is that a hydraulic backfill delivery system would have to be installed and bulk heads would have to be built after each slice of ore is taken. In this version of room and pillar, it is important to have multiple headings available to mine so that the backfill has time to consolidate without interrupting ore production. Mining is usually conducted in an overhand style where the fill acts as the working floor but in some cases, an underhand method is used. Mining can continue with classic room and pillar with horizontal blasting on every level, or rooms can be blasted vertically starting on the second pass. In vertical blasting, the backfill is poured to fill the stope high enough for the drills to operate on the next pass. The disadvantage of this method is that mines may have to purchase two types of drills since horizontal drilling is required in the first pass and every subsequent pass would then use vertical drilling. The image shows typical post pillar development.

Step

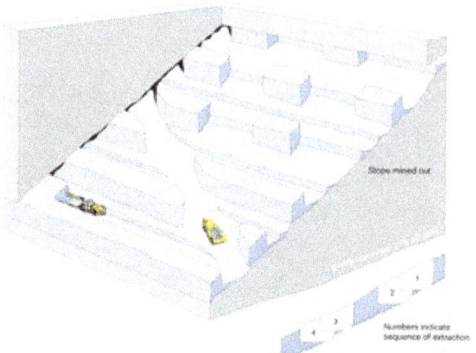

Development in step room and pillar.

Step room and pillar is found in orebodies dipping from 15 to 30 degrees. Orebody thickness is typically quite small, ranging from 2 to 5m. It is essentially an adaptation of 'classic' room and pillar with the orebody being developed in a series of horizontal 'steps'. Haulage ramps are specially design diagonally against the dip of the orebody at shallow enough slopes to utilize trackless equipment. Mining advances downward along the step room angle with each step having a relatively flat production floor.

The main development feature in step room and pillar is the development of an in-clined haulage way within the ore body. The stopes are driven off this haulage way in an opposite direction so that the slope of the stope and the haulage way is appropriate for trackless equipment. Once an initial stope has been developed, the subsequent stopes can be slashed out sideways down the dip, following the slope of the floor. The process of taking adjacent stopes is repeated until a back reaches its maximum span. At this point, pillars are left in place for support. The image shows typical step room and pillar development.

Steep

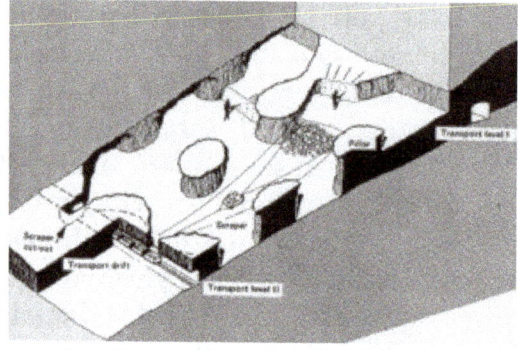

Development in steep room and pillar.

Steep room and pillar, similar to step room and pillar, is found in orebodies dipping from 15 to 30 degrees with limited thickness. It differs in that the working floor matches the orebody inclination. Because of this, trackless equipment is not able to be used and mining is considerably less efficient. This mining method is very seldom used in practice, especially in modern mine systems where significant improvements have been made to mechanized equipment.

This version of room and pillar is less mechanized and requires the use of jacklegs for drilling and slushers for mucking since the slope is too steep for mobile equipment. The first step in development is to drive a horizontal transportation drift across the strike of the orebody. There are cut-outs in this drift that allow the slushers to accumulate the ore in specific locations where it can then be loaded and taken to surface. The pillar shape is usually rounded so that the slusher can efficiently muck the ore out of the stope without being caught up on a square corner or leave excess ore in front or behind a pillar. The image shows typical steep room and pillar development.

Use with Other Methods

Room and pillar is a versatile mining method and can be applied to a diversity of deposits. Often times, variations of room and pillar can be used alongside or in conjunction with other mining methods. In non-homogeneous deposits, room and pillar can be used in flat, thick portions of the orebody meanwhile other methods are being used to access steep, narrow portions. Also, in the case of post pillar mining, room and pillar can be used with cut and fill in an integrated system. Below are a few examples of room and pillar being used with other methods.

Room and Pillar with Sublevel Longhole Stoping

In very thick deposits, sometimes it is best to mine the room and pillar with two levels. Otherwise, development becomes difficult with large room heights. To do this, mines can opt to use sublevel longhole stoping to drill vertically between levels, leaving behind pillars as the primary structural support. This system was used at the Denison Mines at Elliot Lake.

Room and Pillar with Modified Shrinkage Stoping

There are some mines that use modified shrinkage stoping after first-pass room and pillar mining. In low recovery room and pillar mines of sufficient thickness (>15m), "scram drifts and finger raises are developed and the ore is blasted with flat longholes, drilled from a raise with the ore going to what is termed a "modified shrinkage stope". This modified system was used at the Ambrosia Lake uranium district.

Room and Pillar with Cut and Fill

Possibly the most used method used in conjunction with room and pillar is cut and fill. This integrated system is used when backfill is necessitated for pillar structural support.

The most common example of this is the multi-pass systems of post pillar mining where pillars would be too high to support the openings without backfill support. Room and pillar can also be used in very weak deposits where rooms are completely backfilled and pillars are extracted from sublevels beneath.

Drilling and Blasting

Ore production in room and pillar mining uses the same drilling and blasting techniques used in drifting. In this process of full face slicing, the drift dimensions are equal to the width and height of the stope. In cases where the ore thickness dictates that the stope must be mined in multiple benches, vertical drilling can take after the first slice has been taken.

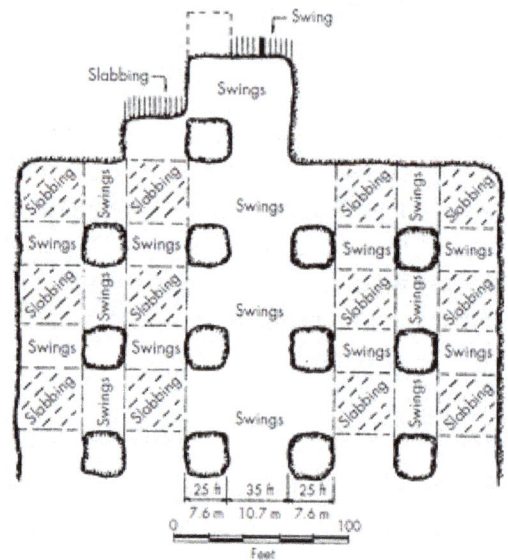

Plan view of a stope showing advancement by swings and slabbing.

The initial advance in room and pillar mining uses a cut pattern when there is only one free face open. The burn cut is the most common drilling pattern in metal mines. This refers to a group of holes that are parallel and centrally located at the face and are detonated on the first few delays. This cut provides relief to the remaining holes allowing them to break. This type of blasting is known as a swing when there is only one free face available. The next type of blasting used in room and pillar is known as slabbing. This type of round is used once a free face has been established so that there is a group of drill holes parallel to an open face. This free face allows the fragmentation of the rock to be the same as a swing with less explosives which leads to lower costs. It is important that the drilling and blasting engineers carefully monitor and plan out the cuts so that the number of slab rounds can be maximized. About half of the rock can be broken using slabbing. The image shows a plan view, showing advancement by swings and slabbing.

Material Handling

In most room and pillar mining operations, ore is usually handled with rubber tire equipment. Due to the nature of this mining method , material handling equipment must be mobile and cannot be fixed in place since stopes are continually developing in different directions. Scoops are used to muck the ore and move it short distances but trucks are typically used the majority of transportation. Depending on the depth and size of the deposit, the mine may use shaft conveyance or ramp haulage. Ore movement from the stope to the shaft or ramp is mostly horizontal, so the practice of using ore passes has very limited applications.

Equipment

Drilling

Drilling in room and pillar is generally carried out with two or three boom jumbos equipped with hydraulic drills. Drill sizes vary by operation but it is usually preferential to use the largest jumbo that is feasible since drilling plays such a key role in the mining cycle. The image shows a picture of a 2-boom jumbo. Drilling is continuously taking place at different faces so it is important that the drills are mobile so that they can move to different areas quickly and efficiently without blocking main haulage drifts. In some cases, crawler mounted long hole drills may be used for vertical drilling, although the vertical drills may provide higher production. Many mines find it more practical to just carry out horizontal drilling so that additional equipment does not need to be purchased.

Atlas Copco 2-boom jumbo.

Mucking and Material Handling

Room and pillar mining is a highly mechanized mining method and makes use of trackless equipment. This results in a very mobile and flexible equipment fleet. Loading of the ore and waste is usually carried out with a load haul dump machine (LHD) which can then tram the ore economically for about 500 feet . LHDs range from 1 tonne to 25 tonne capacities . In cases where the ore needs to be moved over a longer distance, LHDs will load low profile underground trucks. These trucks have capacities of up to 60 tonnes and can economically transport ore over long distances. The use of slushers and dozers are another material handling alternative when rubber tire vehicles are not practical.

In addition to the main production equipment, bolting and scaling rigs are necessary for room and pillar mining. With this mining method comes a very large back that needs to be maintained since the entire stope is open to personnel. These pieces of equipment are an essential part to the mining cycle and to the safety of the workers.

Pillar Recovery

As described previously in the planning (hyperlink) section, the mining engineer must acknowledge pillar extraction in the initial planning phase. Failure to do so can significantly affect the economic success of a mine, as pillar recovery can considerably increase reserves. Oftentimes, mines take too much ore during first-pass mining to a point where the mine cannot support itself, requiring extensive secondary support just to keep the mine in production. When pillars are left too small, convergence can accelerate uncontrollably requiring massive reinforcement or backfill. This reduces the opportunity of pillar extraction and noticeably increases costs. It is usually advisable to take a less aggressive first-pass, and then decide on the second pass how aggressive to be, depending on the stress conditions. This way, support costs are reduced while achieving the same, if not better recovery. There are four main pillar extraction methods depending on the variation of room and pillar being used and the orebody characteristics.

- High grade material can be slabbed off from pillars during a retreat of the mine.

- Pillars can be completely removed in a retreat.

- Select valuable pillars can be completely removed.

- Massive backfill can be used to support the mine, while pillars are typically mine from a sublevel beneath.

Another pseudo pillar recovery method, which is rarely used in practice, is utilized in step room and pillar mining where the hanging wall is caved in moving along strike after first-pass mining.

Advantages and Disadvantages

Advantages

- Flexible – Can utilize multiple faces, and therefore be selective of production and grade. Especially advantageous with base metals that have cyclical price cycles.

- Highly Mechanized – Not very strenuous on the workforce. Allows for high efficiency and productivity.

- Easy Maintenance – Usually utilizes mobile, trackless equipment which is easy to transport in and out of maintenance areas. Equipment can also be transferred easily between levels.

- Low Operating Costs – Largely due the mechanization and productivity, operating costs are usually considerably lower than most underground mining methods.

- Low Development Costs – Most of the development work takes place within the orebody, which means ore production and development work are carried out simultaneously.

- Good Working Conditions – Work takes place in large open stopes with good footing. Room and pillar does not force workers to have to go into confined stopes and stand on top of broken muck.

Disadvantages

- Roof Maintenance – A large portion of the roof is exposed making monitoring and maintenance very time-consuming and costly. The roof can often require high lift equipment for appropriate inspection.

- High Capital Costs – Initial infrastructure and equipment fleet can be expensive, although total costs are typically cheaper in the long run with lower operating costs.

- Low Recovery – Room and Pillar can have one of the lowest recovery rates of any underground mining method. There must be significant reserves left behind to support the mine. Pillars may contain high grade ore which cannot be recovered.

- Lack of flexibility in structural planning – It is difficult to make structural changes part way through production because most of the previously mined out rooms must be supported for the duration of the production life. The stress on a pillar is dependent on the location of other pillars and stress distributions can change drastically with changes in pillar location.

- Traffic Safety Concerns – Due to the large mechanised fleet of equipment required for room and pillar, many piece of equipment must work in close proximity. Traffic accidents and safety of the workers can be an issue.

Sublevel Stoping

The sublevel stoping mining method is usually applied to a relatively steeply dipping, competent ore body, surrounded by competent wall rock. Ore is produced by drilling and blasting longholes, which can range from 50 mm (2 in.) to 200 mm (7% in.) diam, with lengths up to 90 m (300 ft). Longholes can be inclined in any direction, but the ring or pattern usually forms a plane, and the holes are blasted as a unit. Recently developed mobile drilling and loading machinery, as well as new explosives products, blasting techniques, and cemented sand and rock fill have made sublevel stoping a highly efficient and versatile mining method. When designing a sublevel stoping production sys- tem, it should be kept in mind that production rates from conventional sublevel stopes vary widely through- out the life of the stope. Early production is at a low rate, coming only from the drawpoints near the slot, but increases as new drawpoints are reached by the stope face. As the stope nears completion, again, fewer drawpoints are productive. Enough drawpoints must be available at any time to provide required production. Drawpoint availability should be compared to equipment availability; plan for more drawpoints than are needed at any one time. Accurate, realistic scheduling is essential to smooth production rates. Also, initial recovery of ore in a stope/pillar block is normally from 35% to 50% in sublevel stoping. Planning of pillar recovery, representing the majority of ore tonnage in a production block, must be done during early mine planning. Since much of the development already done for primary stoping (access for drilling, drawpoints, and haulageways), can be used for pillar recovery, early production from pillars is highly desirable. The following description of components of the system is an attempt to highlight some of the most important features and requirements of mechanized sublevel stoping methods. Similar comments would apply to the use of older equipment (column-and-arm drill setups, slushers, etc.) in similar methods. As in any good mining system, maximum economic recovery of the resource in the ground is the primary consideration.

Stope Design Characteristics

Length and Width

The following are some of the factors which affect sublevel open stope length and width dimensions: ore body geometry, principal stress directions, competence of stope back, optimum drill pattern, and drilling drift layout. In new mines initial stope layout design may occur before the ore body is actually intersected by mine workings. Stope dimensioning is a critical decision, and assistance from as many knowledgeable people as possible at this stage is essential. Operators with past experience in similar ore bodies, rock mechanics experts, and others with mine design experience should participate at this stage of stope planning.

Height

The following are some of the factors which must be considered in determining stope height: competence of stope pillar and stope/fill walls; slenderness ratio of adjacent pillars; ore body dip; ore body thickness; hole depth capability of the drilling machine; fragmentation characteristics of the ore; and level intervals in existing mines. In competent ground, drill-hole length and accuracy are the most important determinants of stoping height. Frequently entire drilling sublevels can be eliminated because of the depth capability of sophisticated drilling equipment, resulting in significant development cost savings.

Drawpoint Location and Design

Some of the most important considerations of a good drawpoint system are optimum spacing of draw- points, within the constraints of stope dimensions, for uniform drawdown and maximum recovery; excavations designed for stability for the life of the ore block to be drawn-primary stope ore as well as subsequent pillar ore; floor or roadway design including type of surface, reinforcing, grade for water runoff; orientation with respect to the main haulageway, for optimum loader maneuverability and ground stability at the inter- section; and length, to allow articulated front-end loaders to work in a straight configuration. Careful drawpoint design and construction are keys to successful production. Extra care in development, such as smooth wall blasting, rockbolts or grouted rebar, wire mesh, and shotcrete usually will ensure long draw- point life. Human exposure during production loading is of longer duration than during development or production drilling, and consequently preparation of draw- points is easily justified, particularly when pillar ore can be drawn through the same drawpoints. Secondary blasting of boulders can weaken drawpoints, also justifying good ground control techniques. A smooth draw- point floor of poured, reinforced concrete, on a grade of +3% or +4% toward the ore pile facilitates water flow out of the drawpoint, and ease of loader bucket penetration into the muck pile. Slot Raising, Slotting A slot or other space for rock expansion is necessary in conventional sublevel stoping where vertical rings or rows of holes are blasted. The slot can be started at a slot raise driven by conventional raising methods, raise boring, drop raising (predrilling and blasting a raise from the top, using small diameter-less than 200-mm (7%-in.)-holes for relief), or crater blasting (similar to drop raising, but without relief holes). The slot usually extends from the extraction level to the back of the stope.

Block Caving

Block caving is an underground hard rock mining method that involves undermining an ore body, allowing it to progressively collapse under its own weight. It is the underground version of open pit mining.

In block caving, a large section of rock is undercut, creating an artificial cavern that fills with its own rubble as it collapses. This broken ore falls into a pre-constructed series of funnels and access tunnels underneath the broken ore mass. These mineworks are sheltered from the collapsing ore inside bunker-like mass of rock, and miners extract it continuously from here. The collapse progresses upward through the ore body, eventually causing large areas of the surface to subside into sinkholes.

Block caving is used throughout the world. As future mines access deeper and lower-grade ores, this method is likely to be used more widely. Block caving is attractive because it permits very large volumes of ore to be extracted relatively cheaply, increasing production and making lower grade ores economical to mine. Although still more costly than surface mining due to the inherent difficulty of underground operations, block caving is the only underground mining technique that achieves production rates equivalent to surface mining. However, it is only suitable for ore bodies with large horizontal and vertical extents.

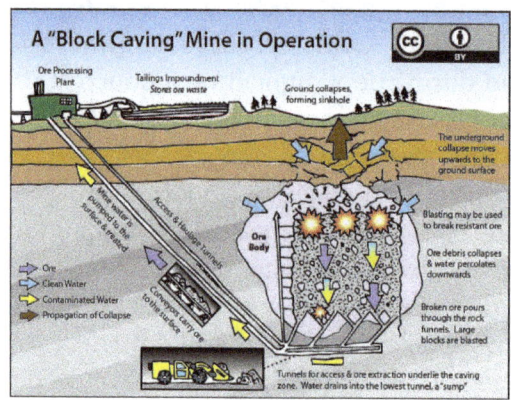

Block Caving Process

Block caving initially involves an elaborate pre-construction process. First, access shafts, must be drilled to a level below the ore. Then horizontal tunnels, known as haulage tunnels, are dug benearth the ore body. From this network of haulage tunnels, a series of upward-sloping tunnels ("raises") are dug. Depending on the integrity of the rock, additional tunnels are dug higher in the rock at the "undercut" level to provide access for blasting to initiate the caving process. Blasting is used to form the "undercut", the extensive horizontal surface from which the ore collapses. Directly below the undercut, large rock funnels ("drawbells") are excavated by blasting upward from the raises. The mouths of these funnels abut one another, forming a continuous plane of funnel mouths where they contact the undercut. Once all the raises, funnels, and undercut are constructed, the main ore body is ready for blasting to shatter it into small pieces.

Ore debris collapses from the undercut and pours through the funnels and into the raises, where special front-end loaders collect it. In theory, no additional blasting or construction is henceforth required. As the funnels empty, the broken ore continues to drop away from the "roof" above the undercut. Unsupported, the ore roof collapses, feeding more material into the funnels. This process progresses until the ore body is exhausted. In practice, the ore body itself usually requires continued blasting to reduce it to manageable size pieces.

Even with extensive blasting of the ore body, large blocks often fall from the ore roof, and block the funnels or raises. Additional chambers, called "grizzlies", are incorporated into raises to trap these blocks and allow them to be broken up.

Hazards and Side Effects

Block caving creates large subsidence features on the surface, such as sinkholes. As a result of removing large continuous masses of rock, the overlying ground surface inevitably collapses to fill the void. Water contamination is a more serious concern. Both surface subsidence features and underground structural changes heavily alter water drainage patterns and groundwater flow. Water is captured by the subsidence features, and percolates downward into the fractured waste rock and remnant ore. There, it can absorb heavy metals and metalloids or react with disrupted sulfide rocks to form acid ("acid mine drainage"). Contaminated water from the mine may enter the groundwater system, pollute aquifers, and resurface to cause problems elsewhere.

From the miner's perspective, however, the primary working hazard of block caving is that the undermined ore body may not collapse continuously. If the underground roof stabilizes, and ore is continuously removed from below, a large void may form. The roof may then collapse catastrophically, producing a destructive surge of air known as a windblast. Windblasts alone can be lethally hazardous, as was seen when a block caving collapse produced a windblast that killed 4 Australian miners in 1999. Large boulders can also cause hang-ups in the drawbells and raises, and can be dangerous to break up, requiring hydraulic breaking or blasting.

Block Caving and Pebble Mine

Block caving has been proposed for a section of the proposed Pebble Mine in southern Alaska to excavate a portion of one of the world's largest gold, copper, and molybdenum deposits. Part of the Pebble deposit is buried roughly 1,000 to 7,000 feet underground. This section is too deep for open pit mining, and the developers have proposed to mine it by block caving.

The Pebble prospect sits at the headwaters of active salmon streams, and the potential for water contamination by block caving has been raised as a major concern. Although the Pebble area is not inhabited, a large subsidence feature at Pebble East could transform the area into a lake. Water contamination could have severe environmental effects if it reached the surface, either as contaminated springs within the lake or at downstream locations, due to acid mine drainage.

Borehole Mining

Borehole mining (BHM) is a remote controlled method of underground mining used to mine a broad range of natural resources and industrial materials such as uranium, iron ore, quartz sand, gravel, coal, gold, diamonds, and amber. A borehole tool comprised of two pipes—one that delivers a stream of high pressure water, and another that delivers slurry back up to the surface—is used. Borehole mining technology is also used in exploration, oil, gas and water stimulation, in-situ leaching, and in the construction of underground storage and drainage systems.

Process

Borehole mining is a fairly new unobtrusive mining approach developed in the 20th century. The process begins with lowering a borehole mining tool mounted on a drill rig tower into a borehole. The drill rig tower provides the tool with the ability to rotate and move up and down along the axis of the borehole. As the borehole mining tool is lowered to a required depth inside the hole, high-pressure water is pumped down and emerges as a high-powered jet stream through a hydromonitor nozzle located near the tool's bottom. The jet stream is so powerful it actually cuts and breaks apart rock mass. Water is added to the broken rock, forming a slurry.

Another portion of water comes up through another part of the tool called a hydro elevator, which is comprised of a jet pump and educator. Together, these components produce a vacuum that sucks and then pumps the slurry back up towards the surface. The slurry of rock debris and water is then separated in a storage tank. Water is then pumped back down to close off the tool's water system.

As a process, borehole mining provides many advantages as compared to conventional mining practices. Areas considered too dangerous to mine using other methods can be accessed with borehole mining equipment. As a process, borehole mining is also extremely mobile and equipment can be transported from site to site. The technique can be implemented on an open-pit mine floor, land surface, underground mine, or on a floating platform or vessel through boreholes that have been predrilled. The cost of borehole mining in comparison to other underground methods is less expensive as it eliminates the need to remove overburden, construct shafts and tunnels, and provide ventilation, de-watering, and transportation of workers underground. The greatest benefit of borehole mining, however, is that it is less invasive on the environment than other methods.

Slope Mining

With fast progress of modern industry, the demand for products manufactured from metal ores has gone up literally. In order to meet this demand, the aware-ness of uninterrupted mining, which can sharply elevate stope productivity and cut mining costs, has become a big trend in the development of mining techniques in underground mines. Classified under the category of main types of underground mines, slope mining is quite similar to shaft mine with the exception of that the coal is closer to the surface and passage to the coal is accessed by a burrow that is dug on slant. Despite the fact that, the actual mine tends to not be very deep, with this mining technique, the coal stratum have a tendency to be very deep with a parallel position to the ground.

With the help of conveyor belt system or track haulage, the coal is transported up to the surface in the slope tunnel. Moreover, electric hoist or a trolley locomotive and steel rope comes into play, if in case the tunnel grade is steep. Based on the depth of the coal seam and the surrounding terrain, the decision of what type of mine to construct is de-cided. More or less all underground mines are not as much of 1,000 feet deep, but some mines reach depths of about 2,000 feet. As a rule, slope mines begin in a valley bottom, and a tunnel slopes down to the coal to be mined.

Shaft Mining

Shaft mining—also referred to as shaft sinking—is a type of mining process used to vertically gain access to an underground mining facility. There are many different components which make up the shaft, all of which play a very important role in the mining process. The entrance to a shaft can go by different names, depending on whether or not the entrance is above or below ground. If it is above ground, it is com-monly referred to as the shaft or portal; if the entrance is underground, it is known as a winze. Winzes, however, are only used in deep shaft mining for connecting lower parts of the mine.

The vertical central shaft of a shaft mine is known as a service cage and is typically used for transporting personnel. Much like a tree, the service cage will have multiple branches extending off of it. These branches go by a variety of names, such as levels, drifts or galleries. The area in which a level will meet with the service cage is known as the shaft station or inset. Each level is perpendicular to the service cage, allowing a horizontal access route to an ore body.

In most shaft mining practices, the shaft will be split into different sections. Each section is vertical, as they run parallel to the service cage. In most shaft mines, most of these sections are used for lifting purposes. The service cage itself usually contains a large elevator, which is used to transport mining personnel up and down the shaft to different levels. The sections are usually rectangular in shape and are lined with either timber or concrete.

Outside the service cage are small shafts known as skips. Shaft mining often requires the use of at least one skip, as they are used to transport ore to the surface. These are essentially smaller versions of the service cage and do not normally carry personnel. Skips can also be used for other necessities, such as pipelines for water and fuel, along with ventilation systems. These are all very essential parts of shaft mining, as water, fuel and air play an extremely important role in keeping the mine—and its workers—fully operational.

On the surface, a head frame is used to winch lifts up and down the shaft. This is done through a hoist motor, which is usually connected to a sheave wheel. Head frames were once constructed from timber, but demands for strength and reliability has resulted in steel and concrete framing. Another role the head frame plays in shaft mining is providing a storage area for ores.

References

- What-is-surface-mining: convergencetraining.com, Retrieved 14 March 2018
- Strip-mining, technology: britannica.com, Retrieved 11 May 2018

- Mountaintop-mining-research-research, water-research: epa.gov, Retrieved 28 April 2018

- 6-pros-and-cons-of-mountaintop-removal: flowpsychology.com, Retrieved 08 July 2018

- What-is-underground-mining: wisegeek.com, Retrieved 17 April 2018

- Hardrock-mining: greatmining.com, Retrieved 28 May 2018

- Block-caving-underground-mining-method: groundtruthtrekking.org, Retrieved 14 May 2018

- What-is-shaft-mining: wisegeek.com, Retrieved 08 July 2018

Chapter 3

Mining Equipments

Heavy machinery is employed for exploring and developing sites, breaking and removing rocks, removing and stockpiling overburden or for processing the ore. Some of the mining equipment commonly used are power shovel, LHD, dragline excavator, gold dredge, crusher, etc. which are vital for drilling and blasting rocks. The topics elaborated in this chapter on such mining equipment and the processes of drilling and blasting, rock blasting, etc. will add valuable insights into the technical aspects of mining.

Drilling and Blasting

Drill and blast method is mostly used method for the excavation throughout the world. The method can be used in all types of rocks and the initial cost is lower than the mechanical method like TBM. This tunneling method involves the use of explosives. Compared with bored tunneling by Tunnel Boring Machine, blasting generally results in higher duration of vibration levels. The excavation rate is also less than TBM (usually 3 to 5m a day).

The typical cycle of excavation by blasting is performed in the following steps:

- Drilling blast holes and loading them with explosives.

- Detonating the blast, followed by ventilation to remove blast fumes.

- Removal of the blasted rock (mucking).

- Scaling crown and walls to remove loosened pieces of rock.

- Installing initial ground support.

- Advancing rail, ventilation, and utilities.

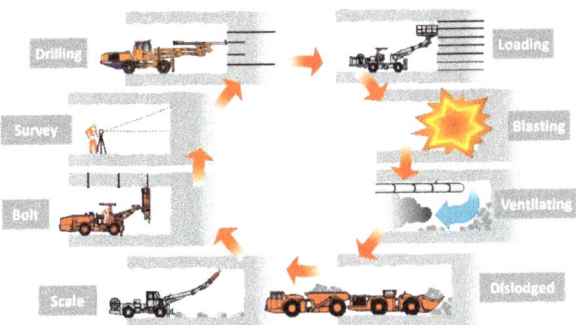

Advantages

- Potential environmental impacts in terms of noise, dust and visual on sensitive receives are significantly reduced and are restricted to those located near the tunnel portal.

- Compared with the cut-and-cover approach, quantity of C&D materials generated would be much reduced.

- Compared with the cut-and-cover approach, disturbance to local traffic and associated environmental impacts would be much reduced.

- Blasting would significantly reduce the duration of vibration, though the vibration level would be higher compared with bored tunneling.

Disadvantages

- Potential hazard associated with establishment of a temporary magazine site for overnight storage of explosives shall be addressed through avoiding populated areas in the site selection process.

Drilling and Blasting Method Sequences

1. Drilling

Before the blasting takes place, the drilling rig bores the drill holes – determined in advance in a blasting plan – in the foremst front wall of the tunnel. The more solid the rock, the more explosives are required.

A jumbo is used to drill holes in the rock face. This one has three drilling arms and an operator tower. It is run by electric cable; a hose brings water to the drills. The drills are pneumatic. That means that the drill bits both hammer and rotate. Broken bits of rock are flushed out by water. These drill holes are 2.4-3.6 meters long.

The first sets are straight holes (parallel cut) located around the edge of the face and in the middle. A second set (V-cut) is angled toward the center. These allow the rock to be blown away from the face into the drift (tunnel).

2. Loading and Blasting

The drill holes are now filled with explosives, detonators are attached to the explosive devices and the individual explosive devices are connected to one another. The holes are blasted in a proper sequence, from the center outward, one after the other. Although more than 100 explosions may be set off, one after the other, the blast sequence is completed in several seconds. The devices should not explode at the same time, but rather one after the other at specified intervals. Only when the blast master has ensured that nobody is left in the danger zone can the explosion be triggered by the blasting machine.

3. Ventilating

The blast causes lots of rock to be flung through the tunnel, dispersing clouds of dust that then mix with the combustion gases of the explosion. So that the miners can resume work in the tunnel, the bad air must be removed from the tunnel. This is done by using so-called air-ducting systems, long steel or plastic pipes, which are attached to the roof of the tunnel and blow fresh air onto the working face. This gives rise to localized excess pressure and the bad air is pushed towards the tunnel exit.

4. Dislodging

Dislodging refers to the stripping away and removal of loose pieces of rock, which were not completely released from the rock during the blasting procedure. This working step is completed by a robust tunnel excavator.

5. Removing rubble

After the loose pieces of rock have been dislodged from the working face, the blasted material – the rubble or spoil – is carried out of the tunnel. The material is either loaded onto dump trucks with wheel loaders and taken from the tunnel to an outside landfill or it is transported from the site of excavation to the landfill on conveyer belts. During the construction of the Brenner Base Tunnel, the transportation of the excavated material mainly takes place automatically using conveyor belts.

6. Securing

The quickly drying shotcrete used for this purpose in particular enables a cavity-free connection of the securing mechanism to the rock. Depending on the type of rock, a variety of securing measures can be implemented: wire mesh, tunnel arches, stakes or so-called bolts, which can be driven into the rock.

The final method for stabilizing rock faces is most commonly rock bolting. A jumbo is used here to first drill holes into the rock. The holes vary from 2.4-6 metres long. Next a steel rod with a wedge threaded on the end is inserted in the hole. When it is in place, the rod is turned so that it pulls out against the wedge, forcing it into the walls of the hole. The outside end of the rod is secured with a steel plate and large nut. Geologists and engineers at a mine determine the spacing and depth of rock bolts required for the conditions at their site.

Under the poorest ground conditions it may be necessary to put steel arches in place to hold up the walls and roof of a tunnel. In other situations a steel mesh may be secured to the walls and roof to prevent other loose materials from falling on workers below.

7. Geological Mapping

The working face is now freely accessible and the geologist has a few minutes to map it. In the process, he determines what type of rock is present and how the rocks lie, i.e. whether they dip in a flat or steep manner, whether they are folded or even broken. He uses a special compass as an aid to measure the angle of incidence and direction of incidence of the rock structures. At the same time, the strength of the rock, the reaction of the rock mass to the excavation process and any mountain water infiltration are also documented. The mapping report created from this – with sketches and photos – serves as the basis for the selection of appropriate supporting measures.

Drilling Patern Design

The drilling pattern ensures the distribution of the explosive in the rock and desired blasting result. Several factors must be taken into account when designing the drilling pattern: rock drillability and blastability, the type of explosives, blast vibration restrictions and accuracy requirements of the blasted wall etc. The basic drilling & blasting factors, and drilling pattern design are discussed below. Since every mining and construction site has its own characteristics, the given drilling patterns should be considered merely as guidelines.

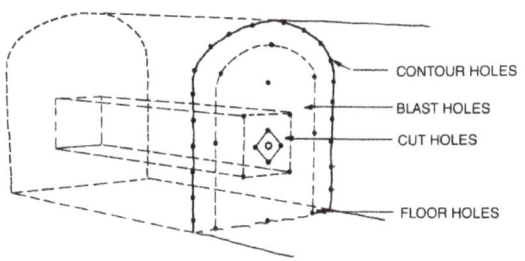

Many mines and excavation sites still plan their drilling patterns manually, but advanced computer programs are available and widely used. Computer programs make it easier to modify the patterns and fairly accurately predict the effects of changes in drilling, charging, loading and production. Computer programs are based on the same design information used in preparing patterns manually.

Drilling pattern design in tunneling and drifting is based on the following factors:

- Tunnel dimensions

- Tunnel geometry

- Hole size

- Final quality requirements

- Geological and rock mechanical conditions

- Explosives availability and means of detonation

- Expected water leaks

- Vibration restrictions

- Drilling equipment

Depending on site conditions, all or some of the above factors are considered important enough to determine the tunnel drilling pattern. Construction sites typically have several variations of drilling patterns to take into account the changing conditions in each tunnel. Drifting in mines is carried out with 5 to 10 drilling patterns for different tunnel sizes (production drifters, haulage drifters, draw points, ramps etc.) The pattern is finalized at the drilling site. Tunnel blasting differs from bench blasting in that tunnels have only one free surface available when blasting starts. This restricts round length, and the volume of rock that can be blasted at one time. Similarly, it means that specific drilling and charging increases as the tunnel face area decreases. When designing a drilling pattern in tunneling, the main goal is to ensure the optimum number of correctly placed and accurately drilled holes. This helps to ensure successful charging and blasting, as well as produce accurate and smooth tunnel walls, roof and floor. A drilling pattern optimized in this way is also the most economical and efficient for the given conditions.

Rock Blasting

Rock blasting is done to break rocks so that it may be quarried or to excavate ground for construction purposes. It is the controlled use of explosives mostly in mining, quarrying and civil engineering such as tunnel, dam or road construction. Blasting is one of the major and greatest inventions in the history possibly after discovery of fire and metals which changed the pace of civilization. Dr.Alfred Nobel famous for the Nobel trust and Nobel prizes is known for inventing dynamite. Blasting, explosives and dynamite became synonymous since then with dynamite being the first safest high explosives.

Impact of rock blasting is enormous and currently utilizes many different type of explosives with different compositions and performance properties. Higher velocity explosives are used for relatively hard rock in order to shatter and break the rock, while low velocity explosives are used in soft rocks to generate more gas pressure and a greater heaving effect. The most commonly used explosives in large scale blasting today are ANFO (ammonium nitrates and fuel oil) based blends due to lower cost than dynamite. Worldwide, huge quantity of explosives is being consumed every day for various Mining and Civil engineering needs. This consumption is also related with the breakage mechanism of rocks and a optimised blast design may in-turn lead to huge savings. Understanding the rock mechanics of blasting would help in safe, efficient and economic blast design and rock breakage.

Figure: A photograph showing large scale blasting in a slope

Explosives

Explosives are mixture of chemical compounds which rapidly decompose, instantly releasing large quantity of energy in the form of heated gas at a high pressure. Its basic ingredients are oxydiser, fuel and a sensitizer. Some of the important properties of explosives are, strength, velocity of detonation (how long it takes to chemical reaction to happen and energy released), density, water resistance, sensitivity, fume characteristic and legal permission.

The strength of an explosive is a measure of the work done by a certain weight or volume of explosive. This strength can be expressed in absolute units, or as a ratio relative to a standard explosive. Usually the bulk strength of explosives is related to the strength of ANFO (ammonium nitrate and fuel oil) that is assigned an arbitrary bulk strength of 100. One measure of the strength of an explosive is its velocity of detonation (VOD); the higher the velocity the greater the shattering effect. However, explosive strength, density and degree of confinement are also factors that should be considered in selecting an explosive for a specific purpose.

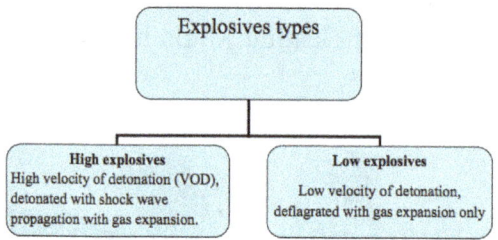

Figure: Type of explosives- High and Low explosives

Examples of high explosives are Nytroglycerine/dynamite, TNT, water gels, special gelatine, slurry, emulsion, ANFO etc. whereas examples for low explosives, gun/ black powder.

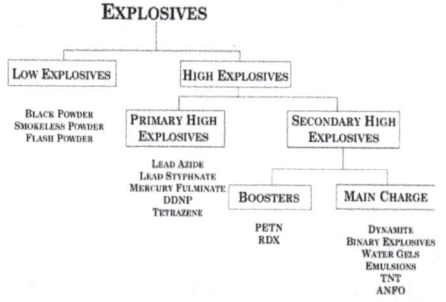

Figure: Different type of explosives used in rock blasting

Detonation of Low and High Explosives

When low explosive is blasted the process of constituent substances is propagated by rapid combustion of particle to particle through the mass of explosive and the effect of explosion is relatively low. A low explosive is fired by ignition or a flame. High explosive always contains an ingredient which is explosive in itself, at least when sensitized by proper means. A high explosive explodes when a violent shock is applied to it with the help of detonator. Here, the process of oxidation doesn't proceed from particle to particle, but is instantaneous and the constituents react with high velocity and produce a shattering effect.

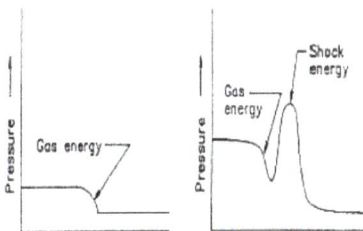

Figure: Detonation of low and high explosives

Blasting Process

When explosive detonates, hole pressure may exceed 20,000-100,000 times than the atmospheric pressure. This also generates stress waves that travel with a velocity of 5000 m/s. The loading front of the stress wave is compressive but is closely followed by a tensile stress responsible for rock fragmentation. A compressive wave reflects when it reaches a exposed rock surface and on reflection becomes a tensile strain pulse. Rocks break much more easily in tension than in compression and fracture progresses backward from the free surface.

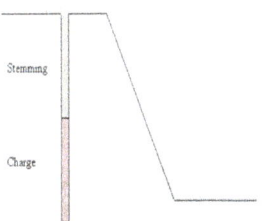
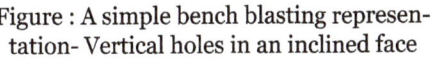

Figure : A simple bench blasting representation- Vertical holes in an inclined face

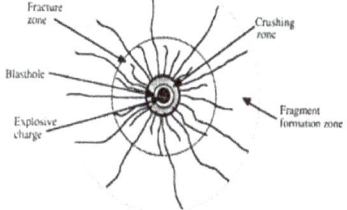

Figure: Illustration of processes occurring in the rock around a blast hole, showing formation of crushing zone, fracture zone and fragment formation zone.

A simple bench blasting representation is shown in figure with vertical holes in an inclined face. There are three basic zones formed during the blasting process, first the pulverized zone, compressive stress zones and followed by radial cracking zone. Schematic illustration of processes occurring in the rock around a blast hole, showing formation of crushing zone, fracture zone and fragment formation zone. Figure shows the reflection of stress wave in the blasting process showing the reflected pulse returning from the free face. A detailed mechanism of rock breakage with explosives with spalling of rock surfaces with returning tensile wave pulse and the expansion of explosive gases with high pressure.

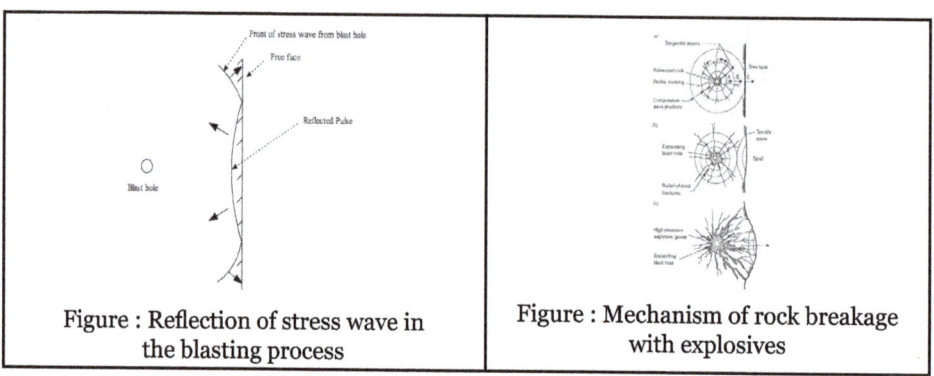

| Figure : Reflection of stress wave in the blasting process | Figure : Mechanism of rock breakage with explosives |

Gunpowder

Gunpowder is a low-explosive mixtures used as propelling charges in guns and as blasting agents in mining. The first such explosive was black powder, which consists of a mixture of saltpetre (potassium nitrate), sulfur, and charcoal. When prepared in roughly the correct proportions (75 percent saltpetre, 14 percent charcoal, and 11 percent sulfur), it burns rapidly when ignited and produces approximately 40 percent gaseous and 60 percent solid products, the latter mostly appearing as whitish smoke. In a confined space such as the breech of a gun, the pent-up gas can be used for propelling a missile such as a bullet or artillery shell. Black powder is relatively insensitive to shock and friction and must be ignited by flame or heat. Though it has largely been supplanted by smokeless powder as a propellant for ammunition in guns, black powder is still widely used for ignition charges, primers, fuses, and blank-fire charges in military ammunition. With varied proportions of ingredients, it is also used in fireworks, time fuses, signals, squibs, and spatting charges for practice bombs.

Dynamite

Background and Raw Materials

Dynamite is a commercial explosive used mainly for demolition and mining. Invented in 1866 by Alfred Bernhard Nobel , it is more accurately described as the packaging of nitroglycerin, a highly poisonous explosive liquid, or other volatile compounds such as sensitized ammonium nitrate. Dynamites can be packed in measured charges, transported easily, and, with the proper detonator, exploded safely. Because a dynamite explosion creates a "cool flame," which is less likely to ignite methane and coal dust mixtures present in mines, dynamites are frequently used in coal mining operations.

Process Design and Facilities

Dynamite manufacture is highly regulated and the process strictly controlled to prevent accidental detonations. The equipment used is specially designed to reduce the exposure of the mixture to heat, compaction forces, or ignition sources. Bearings in the product mixers, for example, are mounted outside of the apparatus frame to prevent contact with the explosive mixture. Buildings and storage areas (called magazines) are constructed at great distances from other structures and with specialized heating, ventilation, and electrical systems. These buildings are "hardened" with bullet-resistant roofs and walls and extensive security systems. Other important precautions include thorough inspection systems which insure correct mixing, grading, packaging, and inventory control. Employees are also highly trained to work with the explosives, and special health precautions are required. Exposure to nitroglycerin commonly produces throbbing headaches, although an immunity to the toxic effects can develop. Interestingly, nitroglycerin is also used in medicine to treat some forms of angina and other ailments. In the body, it acts as a vasodilator and relaxes muscle tissue.

The Manufacturing Process

The process begins with the compound liquid such as nitroglycerin (explosive oil), a "dope" substance, and an antacid. Ethylene glycol dinitrate, composing approximately 25-30% of the explosive oil, is used to depress the freezing point of the nitroglycerin. This allows the dynamite to be safely used at low temperatures. In fact, nitroglycerin in a semi-frozen state with both liquid and solid present is actually more sensitive and unstable than either frozen or liquid state alone. In that semi-solid state, nitroglycerin is extremely dangerous to handle.

Mixing the Oil

The explosive oil is carefully added to a mechanical mixer, where it is absorbed by the "dope," which can be either diatomaceous earth (now no longer used), wood pulp, sawdust, flour, starch, and/or other carbonaceous substances and combinations of substances.

Neutralizing Acidity

Approximately 1% antacid such as calcium carbonate or zinc oxide is added to neutralize any acidity present in the dope. The mixture is monitored carefully and when the correct ingredient level is attained, the mixture is ready for packaging into the various forms. This process produces what is termed "straight dynamite," in which the dope does not contribute to the explosive strength of the dynamite. For example, 40% straight dynamite contains 40% nitroglycerin and 60% dope; 35% straight dynamite contains 35% nitroglycerin and 65% dope. In some cases, sodium nitrate is mixed with the dope, which acts as an oxidizer and gives additional strength to the explosive.

Packaging Dynamite

The appearance of dynamite typically resembles a round cartridge approximately 1.25 inches (3.2 cm) in diameter and 8 inches (20 cm) long. This type is produced by pressing the dynamite mixture into a paper tube sealed with paraffin. The paraffin enclosure protects the dynamite from moisture and, being a combustible hydrocarbon, contributes to the explosive reaction. Dynamite can also exist in many other forms, from smaller sizes of cartridges for specialized demolition work to large 10-inch (25 cm) diameter charges that are used for large strip mining operations. Regulations limit the length of these big charges to 30 inches (76 cm) and the weight to 50 pounds (23 kg). Dynamite is also available as a bag powder and in a gelatinized form for underwater use.

Dynamites are also made using other substances besides nitroglycerin. For instance, replacing a larger portion of the explosive oil with ammonium nitrate can increase the explosive strength of the dynamite. This form of dynamite is referred to as ammonia dynamite.

Quality Control

Accurate dynamite strength measurement and testing by detonation assure safe performance of the explosive. The relative strength of dynamite is graded by comparison to straight dynamite and by the percentage of weight of the explosive oil. For example, ammonia dynamite is compared to straight dynamite and is graded accordingly. Fifty percent ammonia dynamite is equal in explosive strength to 50% straight dynamite. In this instance, the "50%" reflects the strength comparison rather than the explosive content.

After manufacture and batch testing of the dynamite, it is dispensed to the job site under strict transportation and storage regulations.

Application

The following brief example is one of many scenarios for the proper application of dynamite. It must be noted that no one but a certified blasting expert with the correct procedures and equipment should ever attempt to detonate dynamite.

In this example, a rock formation must be blasted to make way for a construction project. The first step in the blasting procedure is to determine the size of the charge by various means, including charts, calculations, and the blaster's experience. Close examination of the affected area and surrounding terrain is made to determine the safe zone. Signs are placed a minimum of 1000 feet (305 m) outside the safe zone to warn the public of the blasting. Radio transmitters are turned off and locked to prevent accidental firing of the electric detonators. The charge is then withdrawn from the magazine and transported to the blast site using closed and secure trucks. The detonators are brought to the job site in a separate vehicle.

The charges are unloaded and placed into the blast holes drilled in the rock formation. They slide into the blast hole by air pressure or by tamping with wooden or plastic rods. The blaster takes great care that the leadwires to the detonators are shorted together until all charges have been placed. This provides a short circuit path for the wiring which prevents accidental ignition. Only the blaster is allowed to make the final electrical connections to the main firing switch.

During this time, a 5-foot (1.5 m) gap in the wiring immediately ahead of the main switch is used as a "lightning gap," another safety practice to eliminate the possibility of static electricity setting off the charges. Once all of the preparation for the blast is complete, a warning horn sounds a one-minute series of blasts prior to the detonation signal. At this time, the final connections to the firing switch are made. At one minute to detonation, a series of short horn blasts are sounded. The blaster then unlocks the main switch and detonates the charges. After the explosion, all electrical circuits to the blasting equipment are once again locked into the safe positions, and the area is inspected for misfired charges and general safety. A prolonged horn blast signals the all clear.

Byproducts/Waste

Explosives manufacture and use contribute some measure of hazardous waste to the environment. Nitroglycerin produces several toxic byproducts such as acids, caustics, and oils contaminated with heavy metals. These must be disposed of properly by neutralization or stabilization and transported to a hazardous waste landfill. The use of explosives creates large amounts of dust and particulate from the explosion, and, in some cases, releases asbestos, lead, and other hazardous materials into the atmosphere. Also, uncontrolled or improperly calculated explosions may rupture nearby tanks and pipelines, releasing their contents into the environment as well.

The Future

Since their development in the 1950s, advanced forms of plastic explosives and shaped charges have replaced dynamite. These explosives are now referred to as blasting agents, since their stability is improved and require a more powerful primer to detonate. One of the most common blasting agent is ANFO, or ammonium nitrate and fuel oil. ANFO is readily available, considerably cheaper than dynamite, and can be mixed on site. However, concrete demolition crews requiring relatively small charges still use dynamite as the blasting agent.

Mining Equipment

Mining equipment is generally used for extraction of minerals, ores and other materials form the earth. In the past, the miners used only simple hand made tools. But thanks to the technological advancements, the miners today can gain many benefits by using modern and state of art mining equipment which can be used for different mining purposes.

Generally, the mining industry is consisted of five major segments. These segments are: coal mining, gas and oil extracting, metal ore mining, non-metal mineral mining and supporting activities (transportation for example). For each mining sector, specific mining equipment is required for performing specific mining tasks.

Various mining equipment is available on the market to suit every need. Some of the most common mining machines that can be seen on every mining site are: excavators, cranes, fork lifts, draglines, rock dusters, tractors, earth movers, water jet pumps, cutting machines drills, loaders, blasting devices, trucks and other mining equipment.

Light mining equipment such as heat sink, LED or laser, magnetic switch and a focusing cone, are also important mining tools. These tools can increase the efficiency and the speed of the mining operations.

Different mining sites require different mining equipment. For example, specific mining equipment is needed for surfaces mining, sub surface mining and for deep underground mining process. Features, restrictions and criteria are important factors that need to be considered when choosing mining equipment. Choosing and using the right mining equipment is essential for increased productivity, safety and money return.

The mining sites are considered as one of the most dangerous work environments. A right piece of mining equipment for a specific mining purpose is essential for preventing accidents and serious injuries. Every miner should be properly trained to operate the mining equipment, and everybody involved in the mining operation should know how to use the mining tools properly. For increased safety, the miners need to be well educated about the safety measures, and to wear safety mining equipment such as: hard hats, protective clothing, protective eye glasses, special shoes, masks, etc.

Drilling Rig

A Drilling Rig is a piece of equipment that is used to create holes or wellbores in the earth's surface. Rigs are massive structures that house all the drilling equipment on board. Some of the major components of a drilling rig are:

- Mud tanks

- Mud pumps

- Mast / Derrick

- Top drive also known as rotary table

- Drill string

- Draw works

- Primary power generation equipment & auxiliary power generation equipment.

They can be massive or small to medium sized structures. The small to medium sized rigs are also called mobile rigs as they are mounted on trucks or trailers and can be easily transferred from one location to another location on wheels. The massive structure rigs can be either onshore rigs or offshore rigs.

Onshore rigs are rigs that perform drilling activities on land, whereas offshore rigs are rigs that perform drilling activities in sea or ocean. Some of the types of drilling rigs used are as follows:

- Land Rigs (onshore rigs)

- Barge Rigs (operates in shallow water)

- Jack up Rigs (operates in water at a depth of 500 ft)

- Semi-Submersible Anchored / Moored Rigs (operates in water up to depths of 10,000 ft)

- Dynamically positioned vessel for Deep or Ultra Deep Water drilling (up to 12,000 ft).

Structure Types

There are many different types of drilling rigs.

There are many different types of drilling rigs. Which rig selected depends on the specific requirments of each drill site. Rigs are generally categorized as onshore (land) or offshore (marine).

Onshore rigs are all similar, and many modern rigs are of the cantilevered mast, or "Jackknife" derrick type. This type of rig allows the derrick to be assembled on the ground, and then raised to the vertical position using power from the draw works, or hoisting system. These structures are made up of prefabricated sections that are moved onto the location by truck, barge, helicopter, etc.

Offshore drilling are divided into two types: Fixed structure types and Floating structure types.

Jack-up Rig

Jack-up Rig is a self-elevating rig and is used for smaller, shallower offshore deposits. The rig's floating platform is towed into position by barges, then lowers its support legs down to the sea floor, raising the rig above the water's surface.

Concrete Offshore Rig

Concrete offshore structures show an excellent performance. They are highly durable, suitable for harsh and arctic environment and can carry heavy topsides. Often offer storage capacities and are very economical for water depths larger than 150m. Gravity type platforms need no additional fixing because of their large foundation dimensions and extremely high weight.

Compliant Tower Rig

Compliant tower rigs are similar to fixed platforms, since both are anchored to the seabed and hold most of their equipment above the surface. Since its design consists of a narrow and flexible tower, it can withstand large lateral forces by sustaining significant lateral deflections.

Barge Rig

Although Barge rigs are not moored to the sea floor, they are explained in this section. They are floating offshore drilling vessels but it is not self-propelled. The drilling equipment is on the barge. It is generally towed to the location and then has its hull filled with water. This type of rig is only used in relatively shallow, swampy areas and are generally capable of drilling in water depths of less than 12ft, or, in the case of a posted barge, perhaps to 20ft.

Submersible Rig

A Submersible rig is a larger version of a posted barge and is capable of water depths of 18ft to 70ft. It has a floating drill unit that includes columns and pontoons that if flooded with water, will cause the pontoons to submerge to a depth that is predetermined.

Semisubmersible Rig

They are the most common type of offshore drilling rigs, combining the advantages of submersible rigs with the ability to drill in deep water. The semisubmersible rig does not rest on the seafloor. This rig is a floating deck supported by submerged pontoons and kept stationary by a series of anchors and mooring lines, and, in some cases, position-keeping propellers. They have a water-depth operating range of 20ft to 2000ft.

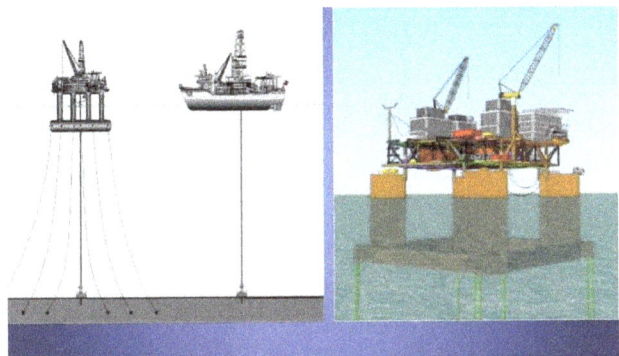

Tension-leg Platform Rig

Tension-leg Platform consists of a floating surface structure held in place by taut, vertical tendons connected to the seafloor. These long, flexible legs allow for significant side to side movement, with little vertical movement. TLP can operate as deep as 7000ft.

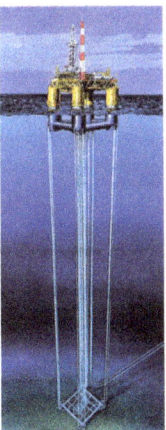

Spar Platform Rig

Spar platforms are among the largest offshore platforms in use. These huge platforms consist of a large cylinder supporting a typical fixed rig platform. The cylinder however does not extend all the way to the seafloor, but instead tethered to the bottom by a series of cables and lines. The large cylinder serves to stabilize the platform in the water, and allows for movement to absorb the force of potential hurricanes.

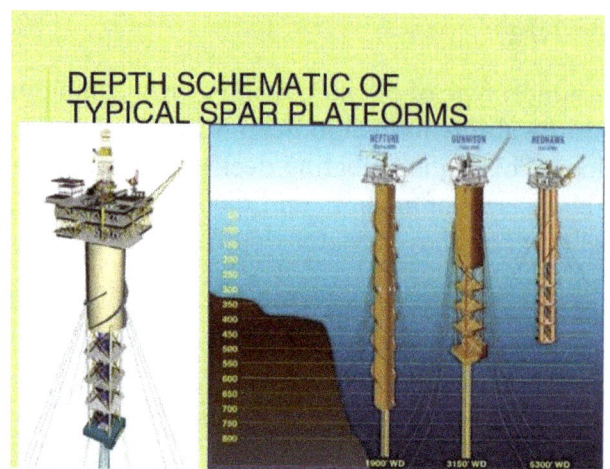

DrillShips

Drillships are most often utilized for extremely deep water drilling at remote locations. A "floater" like the semisubmersible, a drillship must maintain its position at the drilling position by anchors and mooring lines, or by computer-controlled dynamic positioning equipment. Most drillships have greater storage capacity than other types of rigs, allowing efficient operation at remote locations.

Power Shovel

Power shovels are used primarily to excavate earth and load it into trucks or tractor–pulled wagons or onto conveyer belts. They are capable of excavating all classes of earth, except solid rock, without prior loosening. They may be mounted on crawler trucks (crawler-mounted shovels), such shovels have very low travel speed but their wide treads give low soil pressures, which permit them to operate on soft ground. Power shovels may be mounted on rubber tired wheels.

Figure: Illustrates a Crawler-Mounted Shovel

The Size of Power Shovels:

The size of a power shovel is indicated by the size of the dipper, expressed in cubic yards (cubic meters). In measuring the size of the dipper the earth is struck within the contour of the dipper, this is referred to as the struck volume, as distinguished from the heaped volume which the dipper may pick up in loose soil. The bank-measure volume of a dipper will be less than the loose volume. If a 2cu-yd dipper, excavating a soil whose swell is 25%, is able to fill the dipper to its struck volume, the bank-measure volume will be:

$$(Bank\ Volume = \frac{Loose\ Volume}{1+swell} = \frac{2}{1.25} = 1.6\ cu-yd).$$

Power shovels are commonly available in sizes: 3/8, 1/2, 3/4, 1, 1.25, 1.5, 2 and 2.5cu-yd.

The Basic Parts and Operation of a Shovel:

The basic parts of a power shovel include:

1. Mounting

2. Cab

3. Boom

4. Dipper stick

5. Dipper

6. Hoist line

These parts are illustrated in figure.

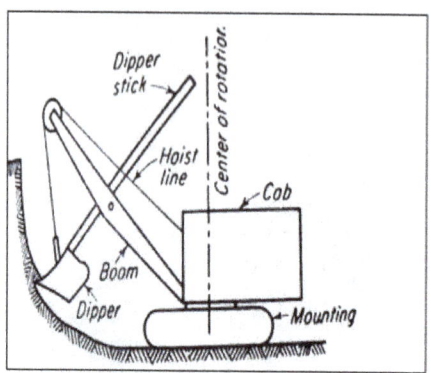

Figure: Basic Parts of a Cable-Operated Power Shovel

With a shovel in the correct position (near the face of the earth to be excavated), the dipper is lowered to the floor of the pit, with the teeth pointing into the face. A crowding force is applied through the shaft and at the same time tension is applied to the hoisting line to pull the dipper up the face of the pit.

If the depth of the face, referred to as the depth of cut, is right the dipper will be filled as it reaches the top of the face. If the depth of the cut is too shallow, it will not be possible to fill the dipper completely without excessive crowding and hoisting tension. This subjects the equipment to excessive strain and reduces the output of the unit.

If the depth of the cut is greater than is required to fill the dipper, it will be necessary to reduce the depth of penetration of the dipper into the face if the full face is to be excavated or to start the excavation above the floor of the pit. The material left near the floor of the pit will be excavated after the upper portion of the face is removed.

LHD

LHD (Load, Haul and Dump) vehicle that is used for the transportation of ore from the underground voids known as stopes, (where the ore is fragmented by blasting), to an ore pass from where ore is transported by gravity to another handling point. The LHD and its operator move back and forth along the mine tunnel, which is typically a few hundred metre slong, hauling the ore. The more repetitions of this cycle that are completed within a shift the higher the production.

LHDs are produced by a number of manufacturers and are available in different models of various sizes using either diesel or electric power. The vehicles typically vary in length from 8 meters to 15 meters, and weigh between 20000 -75000kg and have a transportation capacity of up to 25000kg. The vehicle's body consists of two parts connected together by means of an articulation joint.

The front and rear wheel sets are fixed to remain parallel with the vehicle's body and vehicle steering is achieved by means of hydraulic actuators altering the articulation angle of the vehicle.

An articulated vehicle is preferable in the narrow environment of an underground mine because of its higher maneuverability has also proven that the articulated truck can be modeled by a drift free nonlinear system, with two inputs, namely speed and articulation angle, which is controllable.

A characteristic of multi-axle vehicles is that during cornering the midpoints of their axles tend to follow different trajectories. The difference between these trajectories can be used as a measure of how cumbersome the vehicle is. Figure shows a comparison between the difference in trajectories of the midpoints of the axles (referred to as the of-tracking error) of a car-like vehicle and an articulated vehicle.

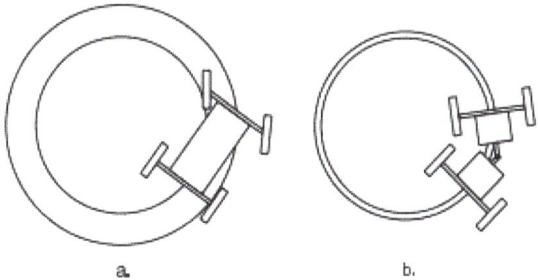

Figure: Comparison between car-like vehicles (a) and an articulated vehicle (b).

The Need for Automation

LHDs are uncomfortable vehicles because they have a low profile for an on-board operator and visibility is made difficult by the fact that the operator usually sits almost at eye level with the top of the vehicle. Several blind spots are created due to the bucket, extinguishers, well covers etc. Stereo vision using digital cameras have been studied to aid the visibility of the operator. The stopes from which the LHD is required to collect ore are hazardous due to the high rock stresses and the likelihood of rock-falls, making them inaccessible to humans. For this reason the LHDs are operated remotely at present, requiring the driver to alight from the vehicle every cycle, which increases the cycle time and the possibility of injury to the driver. In order to prevent this, some mines are now teleremotely operating the LHD vehicles for the entire cycle from above ground. While this has lead to improved safety, these systems unfortunately also lead to a decrease in productivity. The sensory perception of the drivers operating the vehicles from above ground is decreased causing running speeds to be lower, resulting in lower production levels and the additional economic overhead of the infrastructure required for teleoperation.

Several autonomously guided vehicle (AGV) systems have been tested in underground mines. Most of these systems have been based on AGV systems used in industrial environments and are optically guided by means of cameras that follow an optical guide made of a retro-reective stripe or a light emitting rope in the tunnel roof. A commercial high speed underground navigation system called QNavigator is also available which makes use of retro-reective tape mounted on holders on the tunnel walls and a rotating laser scanner. The angle of the rotating head of the scanner is recorded when the beam is reected back into the scanner. The measured angles together with a map of the reector positions are used to determine the position and heading of the vehicle in order to navigate.

Dragline Excavator

Draglines are one of the largest machines used in surface mining. Each dragline is a self-contained excavation system that is able to remove and dispose of the overburden without the assistance of other mining machines. In surface mining, draglines are employed in excavating a long strip of overburden sitting above the target mineral reserve. The typical geometric structure of a surface coal mine without the dragline. A strip of overburden is removed and dumped to the immediately adjacent strip where the coal has already been extracted. A large pile of material is formed by the dumped material and extended on the adjacent strip, which is called the spoil pile. The excavation of a strip of overburden is scheduled in a sequence of blocks along the strip. Each of the blocks contains a certain amount of overburden and is defined by the strip width, the block length, the overburden depth and the angle of the working surface. A block of overburden is removed while the dragline moves through a sequence of position (positioning). At each position, the dragline performs cyclic operations of digging the material from the block and dumping it onto the spoil pile. Specifically, each dig-dump cycle comprises: filling the bucket by dragging it with the drag rope towards the fairlead; hoisting the loaded bucket with the hoist rope while swinging it to the spoil pile; dumping the material; and swinging and lowering the bucket back to the block to start the next cycle. The drag rope and the hoist rope are con- trolled by separate motors that are located inside the machinery house. A dragline repeats this dig-dump cycle a thousand times a day on average with a cycle time of around one minute.

Currently the dragline operations, including digging, dumping and positioning, are planned and executed by the operators.

Bucket-wheel Excavator

A Bucket Wheel Excavator or BWE is a specialized type of heavy equipment used extensively is surface mining operations. These excavators function in large scale digging operations such as open pit mining as a machine for continuous digging. This type of excavators are different from other excavators for large scale mining like bucket chain excavators because they scoop materials using buckets placed in a continuous pattern within a large wheel. Bucket Wheel Excavators are considered to be one of the largest vehicles ever built by man. Interestingly, Bagger 293, the largest BWE ever constructed, is recognized by the Guinness Book of World Records to be the largest land vehicle ever.

The superstructure is the most important component of a bucket wheel excavator and several other components are attached to it. The bucket wheel is a large wheel of large size with different scoop configurations. These configurations can rotate and are attached to a boom. The cutting wheel picks up materials and transfers them back along its boom. In previous designs, materials were transferred through chutes originating from the buckets. Nowadays, the same function is accomplished by one stationary chute that collects the discharge from all of the buckets. The cutting boom is balanced by a counterweight boom, and the same is cantilevered on the upper or lower end of the superstructure. In case of larger bucket wheel excavators, there are cables that run across the towers, providing support to all three booms.

The movement system of a bucket wheel excavator lies beneath its superstructure. Rails were used for this purpose in the older machines. However, the advanced models use crawlers for a greater flexibility in terms of motion. BWEs also have a vertical axis that allows the superstructure of the machine to rotate and complete its tasks.

Depending on the application intended, bucket wheel excavators are available in different sizes. The compact ones with only 6 meters of length and 50 tons of weight are available for operations that require the movement of 100 fm3/hr of earth. The boom length for the larger models can be as high as 80 meters and their weight can be approximately 13,000 tons. With these machines, you can move up to 12,500 fm3/hr.

Bucket wheel excavators are used in surface mining applications where overburden removal is required on a continuous basis. These jobs were previously accomplished using draglines and rope shovels. However, bucket wheel excavators are now the most reliable alternatives because of their extraordinary efficiency.

Gold Dredge

In the beginning of the Gold Rush, the miners were limited because they could only work the areas that were accessible to hand tools along the banks of the streams and rivers. Their equipment was limited to gold pans, sluice boxes and rockers. As time progressed and as they became more experienced, they realized that the deeper gavels in the riverbeds were often richer than the surface gravel along the banks.

In the early 1900s, several crudely built steam powered dredges were active on some of the northern rivers of California. The divers worked futility on the bottom of the rivers with heavy diving helmets and cumbersome diving suits.

Although history reveals that dredging has been in existence throughout the world for many years, it is just recently that it has reached such a high degree of popularity due to advanced technology in dredging equipment. Dredges of today are lighter, more portable and more efficient than ever.

Two Divers operating a New Keene 4 inch Dredge in shallow water.

A small portable backpack dredge of today can weigh as little as forty pounds and cost around eight hundred dollars. It can process as much gravel as a larger three hundred-pound dredge, some twenty years ago. One of the most exciting features of this type of dredge is that it allows the prospector to penetrate areas that were otherwise impossible to reach with heavier and more cumbersome equipment.

They are also far more efficient than the machines of old. It is not uncommon to see a dredge profitably working the tailings of some of the old mines and tailing dumps.

There are three basic types of dredges on the market today. They include surface dredge, submersible dredging tube and the underwater submersible dredge. The surface dredge is the most popular, efficient and versatile gold recovery machine.

The Surface Dredge

As the heading of this section implies, surface dredge floats on the surface of the water. The material is pumped to the surface through a suction hose into an efficient sluice box that is capable of recovering extremely fine particles of fine gold. The sluice box can either rest on the bank, or float on the surface of the water. Another advantage of the surface dredge is it can easily be operated with or without diving equipment. Marlex plastic floats are normally preferred as they are rugged and stable in rough water and are extremely lightweight.

Medium sized 4" gold dredge

Modern dredges are provided with a Single or "multi stage" recovery systems, such as a the new 3 stage sluice. The multi stage sluice boxes are preferred, because they have a greater capacity to recover a finer grade of gold and black sand concentrates that often hold values. Normally, the smaller size dredges from two to three inches in size are equipped with single sluice boxes, as their primary function is portability and compactness.

Large 8" dredge

Small back pack 2" dredge

The Underwater Dredge

The underwater dredge is the less popular of the dredges available, because it lacks somewhat in its ability to recover as fine of gold as the surface type. It is designed mainly for compactness and portability, but is limited also in its application, as it is cumbersome to handle underwater.

The submersible dredge must be held relatively level while in operation and cannot reach around corners and hard to get at places. It also is not practical to use in shallow water, as it must be completely submerged in order to operate properly. It is physically described as a flared metal or plastic tube with an attached metal elbow at a forty five to sixty degree bend. High pressure water is pumped into the bend, creating a vacuum at the end of the bend. It is powered by a high pressure water pump which is normally located on a float that sets at the surface of the water and is pumped down to the dredge via a high pressure hose. At the end of the flared tube a riffle tray is attached containing a series of gold traps. As the gold bearing gravel is sucked into the dredge the heavier particles, including gold, becomes entrapped into the riffle tray. The lighter non gold bearing particles flow back into the river.

The submersible dredge of today is mainly used for sampling and when a good streak is found, the surface dredge is employed to do a more efficient job of recovery.

Submersible Suction dredge uses power jet and suction hose. Suspended under the water by a float system.

Submersible dredging tube uses a suction nozzle. It is handheld underwater for moving overburden quickly.

Single Sluice Box, a Double Sluice Box, a Triple Sluice Box and the Latest Technology of the 3 Stage Sluice Box

The single sluice box processes all dredged material through a single recovery box. A single box includes a short classifier screen at the entrance, to separate the larger cobbles from the smaller, which are most likely to contain small gold particles. This design is still used in most small dredges, due to their lightweight compact design.

The double or triple sluice also separates and classifies the dredged material at the entrance of the sluice. The smaller heavy material falls into separate sluices for a more selective recovery. When the dredged material is separated by size, it ensures better recovery. Higher velocity water is required to move the larger cobbles through the sluice. Lower speed or velocity is required to recover fine gold in the lower or separate sluices. When the speed is high enough to carry off the larger non-value cobbles through a single sluice box, a loss of fine gold can occur if the material and flow is not separated.

The latest technological design is the new 3 stage sluice box. The new box works similar to the double or triple sluice, classifying the fine material into separate compartments or sluices for processing. The differences are:

1. The material is classified 1/3 of the way down the box allowing the fine gold to fall and settle out of suspension.

2. The material passes over two different classifier screens allowing more complete separation. Experience finer gold recovery, in a lighter and more compact dredge that will out perform any double or triple sluice on the market! This system is extremely easy to operate for quick and easy clean up. It provides quick identification of values in the primary recovery riffle section.

The top edges of the sluice box are rolled for Greater safety and strength. They are equipped with heavy duty latches and a longer rubber damper that is used for more even distribution of material over the recovery area to assist in settling fine gold out of suspension. Currently available only with 4, 5, and 6 inch dredges.

Representation of a 3 Stage Sluice Box

General Operating Instructions

The following information should provide you with a basic understanding of operating a portable dredge. For more complete understanding on this subject, we recommend you read any one of a variety of books available through the Keene Library of Books, such as The Gold Miners Handbook, Dredging for Gold or Advanced Dredging Techniques.

The vacuum on a portable dredge is created by a "venturi principal". A volume of water is pumped through a tapered orifice (jet), by a special designed water pump. A high velocity jet stream is created within the jet tube producing a powerful vacuum. As indicated in the diagram gravel is dredged into the suction hose and is delivered to the sluice box header. As a slurry of water and gravel enters the header box and is spread evenly over a classifier screen. The smaller and heavier particles drop below the classifier screen into an area of less velocity, allowing a slower and more selective classification of values. Often values are recovered and easily observed before they even enter the riffle section. The lighter non bearing values and larger aggregate are returned back into the water. The riffles, or gold traps in the sluice box are best described as "Hungarian Riffles". This type of riffle has proven to be the most efficient gold recovery system. As material flows over the riffles a eddy current is formed between each riffle opening. This force allows the heavier material to settle out of suspension and the lighter, non value bearing material to be washed away. This continuous self cleaning principal allows a dredge to be operated for prolonged periods of time. Normal conditions require a sluice box to be cleaned only once or twice a day.

Priming the Pump

Before starting the engine, the pump must be fully primed. This means the pump must be full of water and all air removed. All jetting pumps provided with our dredges have a mechanical water pump seal. Without the presence of water in the pump, friction could cause a seal to overheat and require replacement. Priming the pump on some of the smaller models is accomplished by thrusting the foot valve back and forth under the surface of the water in a reciprocating motion. This will cause water to become pumped into the foot valve assembly into the pump. A pump is fully primed when water is observed flowing out of the discharge end of the pump. It sometimes may become necessary to hold the discharge hose above the level of the pump to complete the priming operation. The larger dredges that have a rigid foot valve, are easily primed by removing the cap provided on the foot valve and filling, until water overflows. Caution must be exercised to prevent sand from entering the foot valve or intake portion of the pump. Excess amounts of sand could dam age the water pump seal, or pump impeller. It is recommended that the intake portion of the foot valve be placed in a sand free environment underwater, such as a small bucket or pan.

Priming the Suction Hose

Priming the suction hose need not be of concern in most dredging operations, but is important to understand the principal. When the tip of the suction hose is taken out of the water during operation air will to enter the suction system and cause the suction power to cease temporarily, until submerged again. The suction will commence as soon as the air is passed through the system. It is important to ensure that no air leaks occur in the suction system.

Suction System Obstructions

The suction system can become jammed while dredging. This can be caused by dredging an excess of sand, causing the suction hose to load up, or a rock that has become stuck in the suction system. Rock jams generally occur in the jet, or just before entry into the jet. This can easily be cleared by removing the rubber plug located on the front of the header box and thrusting the probe rod through the header box and down through the jet in an effort to strike the obstructed area. It may occasionally be necessary to remove the suction hose to remove an obstruction. Sometimes obstructions can easily become dislodged by back flushing the system. Back flushing a suction system can be accomplished on some models by reversing the flow of the suction hose at the header box, by blocking the flow of the water as it enters the header box. If this is not successful. it may be necessary to locate the blockage in the transparent hose and dislodge it by a striking the obstruction, taking care not to damage the hose.

Solid Content

Care must be exercised to prevent dredging excess amounts of sand. A solid to water balance must be maintained. The solid content being dredged should never exceed 10%. If a suction tip is buried into the sand and not metered properly the solid content could cause the suction hose to become overloaded with solids and suction will cease, this will also cause the sluice box to become overloaded with solid content, resulting in a loss of values.

Sluice Box Adjustment

Most models have a slight adjustment to raise or lower the sluice box. The proper sluice box adjustment can effect the recovery of values. If the sluice does not have enough angle, the sluice box will "load up" causing the riffle openings to fill with unwanted excess material. Too much angle will cause the material to flow too fast, resulting in loss of values, evidenced by the riffles running too clean. The optimum adjustment of a properly working sluice box is evident by only a portion of the riffle visible while operating. A loss of values can also occur if the solid content of the suction discharge is too heavy in solid content. Remember, the solid content should not exceed 10 %. A normal sluice box tilt is approximately 1/2 inch to the running foot. A four foot sluice box should have an approximate tilt of 2".

Cleaning the Sluice Box

Before attempting to clean the sluice box, it should be allowed to run with only water for a few minutes in order to wash out any excess gravel that have accumulated. Either turn engine off, or let run with a slow idle, then remove the classifier screen and replace the wing nut to prevent losing it. Unsnap the riffle latches, fold the riffle tray up, and let rest against the header box, taking care not to let it drop back into place while cleaning. This could result in a potential injury! Place a wide tray, bucket or large gold pan at the end of the sluice, then carefully roll up the riffle matting and wash into the container at the end of the sluice. Rinse any excess gravel that remains in the sluice into container. All material must be removed before replacing the riffle matting, riffle tray and classifier screen.

Engine Speed

Most small engines are throttle controlled. The speed of the engine can be controlled with the use of a lever. Although the rated horsepower is achieved on most small engines at 3600 R.P.M., it may not be necessary to operate the dredge at full speed. Lower speeds conserve engine life and fuel economy. Be sure to read all instructions and especially the engine instructions that are provided with each unit.

Trouble Shooting

If Suction Declines:

1. Check the suction device for an obstruction. An obstruction can be removed by probing the obstructed area with the provided probe rod. I may be necessary to check the suction hose for a visible obstruction. This can be remedied by either back flushing the system or dislodging the obstruction with a gentle blow.

2. Check the pump for loss of prime or blockage. The foot valve may be too close to the surface of the water and air may enter the intake of the pump via a small whirlpool. The pump intake or foot valve screen may be plugged with leaves or moss, restricting flow into the intake of the pump. Check and tighten all clamps to prevent an air leak.

If Priming the Pump Becomes Difficult

1. Check all clamps for an air leak.

2. It may be necessary to check the foot valve for a small leak. This is accomplished by removing the foot valve assembly from the pump and blowing air into the hose portion of the assembly and listening for an air escape. It may be necessary to remove the hose and check the rubber valve for an occurrence of a leak, or for a small obstruction preventing the valve from sealing.

3. If a water pump seal is either defective or damaged, a leak will be evident on the inside portion of the pump around the drive shaft. Often a new pump will leak slightly, until the seal and gasket has become fully seated. This is a common occurrence in most new pumps.

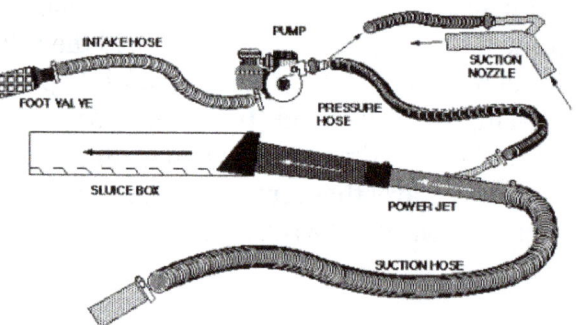

Figure: The Principal of Dredge Operation

Raise Borer

Raise boring is used in underground mining to create a circular hole between two levels within a mine. This method eliminates the need for explosives. The raise boring machine is placed on the level above the two levels which are going to be the focus of the raise boring. Once the top level has been broken through and a pilot hole is placed into to upper level of the two levels you are focusing on, a reaming head is used to create the required tunnel in the exact size that is needed.

Being able to carry out this heavy duty mining activity, without having to use explosives, is the feature of Raise Boring. Blasting can be incredibly dangerous, it can lack precision and it wastes a huge amount of time. When using the raise boring method, the mine on which you are working does not have to be evacuated and the raise boring can be completed during the shift times. This saves time because the mine does not have to be closed while the raise boring is taking place.

Raise boring has become a popular method to use when creating holes between the two levels. The method became popular because it saves time, it saves money and it is the safest way in which you can create a hole.

Methods of Raise Boring

Pilot Drilling

Conventional raise boring begins with the drilling of a pilot hole. It is drilled using a roller bit with sealed bearings, together with hollow drill pipes 1.5 m in length with an international standard thread for high–torque applications. The cuttings are removed from the pilot hole with the aid of water flushing. Introduced through the center of the drill string, the water flows out of the drill bit and up through the annulus between the drill pipes and the hole wall. If required, the pilot drilling can be controlled by using a directional drilling system.

Raise Boring

When the pilot hole breaks through into the lower level, the roller bit is removed and replaced with a reaming head. The reamer is rotated and pulled back toward the drilling unit. The cuttings fall by gravity into the chamber at the bottom of the hole, where they are mucked out using a LHD-type loader.

Raises up to 6 m in diameter and up to 1000 m in length are not uncommon. The raise-boring method is used to produce ventilation shafts, ore passes, manways, penstocks etc.

Blind Boring

When a raise is required but there is no access to the upper level, it has to be bored blind from below, usually without a pre-drilled pilot hole. A special type of head is required for blind boring. It drills the pilot hole and reams out the raise at the same time. The head is rotated and pushed upward. The cuttings fall out of the hole by gravity.

Normal blind raise diameters are from 0.6 to 1.8 m. Since the drill string is under compression during blind boring, special large-diameter stabilizers are needed to support the drill string. The blind-boring method is used to produce so-called slot raises, ore passes and manways.

Horizontal Boring

Horizontal boring is an excellent method in urban construction projects where drilling and blasting is restricted or forbidden and tunnel boring machines (TBMs) are too bulky. First, a horizontal pilot hole is drilled, with the aid of a directional drilling system if necessary. When the pilot bit breaks through, it is removed and replaced with a reaming head. Because the hole is horizontal, the reamer must be equipped with a special cuttings removal system. Typical diameters for horizontal reaming are from 0.6 to 4.5 m. The method is used to drill tunnels for cables, escape routes, sewage etc without disturbing the environment unduly. Horizontal boring requires good rock stability.

Down Boring with a Pre-drilled Pilot Hole

In mines, large fill-holes between 0.6 and 1.8 m in diameter can be bored into stopes using reaming equipment, provided that a pilot hole can be pre-drilled into the stope. The pilot bit and drill string are then removed and a reamer fitted. The reamer is pushed and rotated downward, guided by a nosepiece that follows the pilot hole.

The cuttings fall by gravity down through the pilot hole. Since the drill string is under compression during down boring, special large-diameter stabilizers are needed to support the drill string.

Hoist

Mine hoists are the most widely used pieces of mining equipment in the industry. Mine hoists are responsible for conveying, maneuvering, and transporting mined materials through the mine shaft and among machinery parts. Mine hoists can take a variety of shapes and come in several sizes. It is not uncommon for a single machine to employ several mine hoists in its operative design.

Metals and minerals are mined from a variety of the earth's layers. Mine hoists enable deep mining and narrow shaft mining to take place without having to erode more of the earth than necessary in shaft digging and material excavation. Mine hoists make even the most complicated material excavation possible, enabling mining industry equipment to make great strides in resource extraction. Mining hoists widen the capability of large mining equipment and underground mining equipments efforts by expanding mobility and increasing lifting power to machinery capability.

Mine hoists make the mining process easier, safer, and more efficient for the entire mining industry. Gold mining and coal mining alike are made more easy with mine hoists and hoists technology. Motorized mine hoists make production run at a predictable pace, while manual mine hoists rely upon human power to propel lifting and lowering of material and are therefore harder to predict a standard timing or pace. Mine hoists of all types have their individual benefits. Each of these benefits make mining equipment more efficient for the project at hand. Many companies exist to provide support, maintenance, modification, refurbishment, and fabrication service options to each mining equipment client, which includes mine hoists and mine equipment machinery. Mine hoists can be modified for the specific needs of any project.

Underground mining equipment require heavy machinery mine hoists to gather regulated amounts of material from a specified depth and normally into a shaft. Mining equipment consists of several phases of machinery. Mine hoists may be employed in each type of machine. Mining industry equipment relies upon steady production to ensure optimal profits. Mine hoists can be implemented throughout any of the stages or styles of mining. Mine hoists must be secure for stable production. Repair to these machines can be costly and cause setbacks to production for the entire mining operation.

Types of Mine Hoists

Air or Steam (Anaconda Type) Hoist

This is a rugged, compact and efãcient hoist applicable for use with either steam or air for surface prospects or underground work in larger mines. It is a simple machine with few wearing parts; "V" friction clutch. Air requirements 125 to 150 c.f.m.

Beebe Hand Hoist

This unit easily handles a multitude of hoisting problems. It is easily portable and the large gear reduction makes minimum effort required to lift large loads. An ideal unit for trucks, hand derricks, gin poles, jib cranes and similar applications. Power may be applied to any size unit when desired.

Mine Hoist

One of the main problems in the operation of a mine is the hoisting of ores speedily and without excessive labor costs. Mine Hoists are of rugged construction and time proven design and successfully solve this problem. They keep tonnage moving, without shutdowns or delays which reduces overall operating expense. Frames are sectionalized and of heavy steel construction. Drums and gears are of Meehanite and pinions of forged steel. Heavy duty ball bearings, removable and adjustable, are used on the sizes shown. Post brakes are constructed of heavy cast steel and designed with a large safety factor.

Larger sizes of hoists than those listed can be furnished, as well as hoists with different speciãcations such as drum friction clutches, band brakes, or double drum. Additional data gladly furnished upon request.

Eberhardt Utility Hoist

The Eberhardt Hoist is a small, substantial and simple electric hoist which may be installed any place, indoors or out, and operated by anyone. This hoist requires no attention after it is installed except for oiling about twice a year, yet it is always ready for use. It is built for required rope pulls up to 1100 pounds.

The following are a few of its many useful applications; lifting and lowering liners when relining a ball mill or rod mill; hauling grinding rods, mine timbers, rails and ties; spotting mine cars, and for hoisting or lowering of light loads.

The hoist consists of a high torque (fully-enclosed if used outdoors) electric motor, directcoupled to a worm gear speed reducer, all mounted on a welded steel base. A machine-ãnished drum is mounted on the output shaft of the reducer. The worm in the reducer is machined integral with the shaft and is case hardened, accurately ground and polished. Gears are cut from high-grade gear bronze. Both the worm and gear shafts are mounted on oversize antifriction bearings.

The hoist can be operated by one man from a remote station using a "start-stop-reverse" push button and a reversing magnetic starting switch which is provided with overload relays and low voltage protection. The load is held automatically by a brake on motor armature shaft.

Utility Hoist and Tugger

The hoists illustrated are two of a large variety available for underground duties. The single drum air hoist is extremely useful for such purposes as operating a skip in a winze or for pulling cars on an incline. This hoist can be equipped with column mounting and can be readily installed for the service required. It operates under 30 lbs. per square inch air pressure.

The double drum electric hoist is most suitable, where power is available, for use in connection with scraper mucking and loading, or "slushing" operations. It is also furnished in single and triple drum types and can be driven with electric motor, air motor, gasoline engine or oil engine drive. Capacities range from 750 lbs. to 10,000 lbs. with rope speeds varying from 75 feet to 450 feet per minute on various sizes available.

Mining equipment involves the use of mine hoists in many procedures. Mine hoists can be manual or motorized, and can be used to convey a variety of material. Mining equipments manufacturers have revolutionized mine hoists in recent years. Single drum, double drum, and friction hoists are commonly used in Canadian mining equipment projects. Iron ore mining equipment, placer mining equipment, industrial mining equipment, and underground mining equipments are also examples of machinery which involve the use of mine hoists in their design.

Metals and minerals are mined from a variety of the earth's layers. Mine hoists enable deep mining and narrow shaft mining to take place without having to erode more of the earth than necessary in shaft digging and material excavation. Mine hoists make even the most complicated material excavation possible, enabling mining industry equipment to make great strides in resource extraction. Mining hoists widen the capability of large mining equipment and underground mining equipments efforts by expanding mobility and increasing lifting power to machinery capability.

Mine Hoist & Mining Process

Mine hoists make the mining process easier, safer, and more efficient for the entire mining industry. Gold mining and coal mining alike are made more easy with mine hoists and hoists technology. Motorized mine hoists make production run at a predictable pace, while manual mine hoists rely upon human power to propel lifting and lowering of material and are therefore harder to predict a standard timing or pace. Mine hoists of all types have their individual benefits. Each of these benefits make mining equipment more efficient for the project at hand. Many companies exist to provide support, maintenance, modification, refurbishment, and fabrication service options to each mining equipment client, which includes mine hoists and mine equipment machinery. Mine hoists can be modified for the specific needs of any project.

Underground mining equipment require heavy machinery mine hoists to gather regulated amounts of material from a specified depth and normally into a shaft. Mining equipment consists of several phases of machinery. Mine hoists may be employed in each type of machine. Mining industry equipment relies upon steady production to ensure optimal profits. Mine hoists can be implemented throughout any of the stages or styles of mining. Mine hoists must be secure for stable production. Repair to these machines can be costly and cause setbacks to production for the entire mining operation.

Mine hoists are implemented into all types of industrial mining equipment, and comprise a significant amount of mining equipment repair orders. Canadian manufacturers of mining equipment provide the knowledge and experience from a diverse staff of professionals who regularly supply repair, refurbishment, or maintenance solutions to mine hoists and mining machinery.

Crusher

A crusher is a multi dimensional machine which is designed to reduce large size materials into smaller size materials. Crushers may be used to reduce the size, or change the form of waste materials so they can be more easily disposed of or recycled, or to reduce the size of a solid mix of raw materials (as in the case of ore), so that pieces of different composition can be differentiated for separation.

Crushers are normally low speed machines that are designed for breaking large lumps of ores and stones, even having a size with a diameter of over one and half meter. The purpose of crusher is to reduce the size of the materials for making them usable in construction or industrial use, or for extraction of valuable minerals trapped within a ore matrix.

Crushing is the process of transferring a force amplified by mechanical advantage through a material made of molecules that bond together more strongly, and resist deformation more, than those in the material being crushed do. Crushing devices hold material between two parallel or tangent solid surfaces, and apply sufficient force to bring the surfaces together to generate enough energy within the material being crushed so that its molecules separate from (fracturing), or change alignment in relation to (deformation), each other.

There are four basic ways to reduce a material, namely (i) impact, (ii) attrition, (iii) shear, and (iv) compression. Most crushers employ a combination of all these crushing methods.

- Impact – In crushing terminology, impact refers to the sharp, instantaneous collision of one moving object against another. Both objects may be moving, or one object may be motionless. There are two variations of impact, namely (i) gravity impact, and (ii) dynamic impact. Material dropped onto a hard surface such as a steel plate is an example of gravity impact. Gravity impact is most often used when it is necessary to separate two materials which have relatively different friability. The more friable material is broken, while the less friable material remains unbroken. Separation can then be done by screening. Material dropping in front of a moving hammer (both objects in motion), illustrates dynamic impact. When crushed by gravity impact, the free-falling material is momentarily stopped by the stationary object. But when crushed by dynamic impact, the material is unsupported and the force of impact accelerates movement of the reduced particles toward breaker blocks and/or other hammers. Dynamic impact has definite advantages for the reduction of many materials.

- Attrition – It is a term applied to the reduction of materials by scrubbing it between two hard surfaces. Hammer mills operate with close clearances between the hammers and the screen bars and materials reduce by attrition combined with shear and impact reduction. Though attrition consumes more power and exacts heavier wear on hammers and screen bars, it is practical method for crushing the less abrasive materials such as limestone and coal.

- Shear – It consists of a trimming or cleaving action rather than the rubbing action associated with attrition. Shear is usually combined with other methods. For example, single roll crushers employ shear together with impact and compression. Shear crushing is normally called for under the conditions when material is somewhat friable or when a relatively coarse product is desired. It is usually employed for primary crushing with a reduction ratio of 6 to 1.

- Compression – As the name implies, crushing by compression is done between two surfaces, with the work being done by one or both surfaces. Jaw crushers using this method of compression are suitable for reducing extremely hard and abrasive materials. However, some jaw crushers employ attrition as well as compression and are not as suitable for abrasive materials since the rubbing action accentuates the wear on crushing surfaces. As a mechanical reduction method, compression is to be used if the material is hard and tough, if the material is abrasive, if the material is not sticky, and where the finished product is to be relatively coarse.

The above four methods for the size reduction of materials are shown in figure.

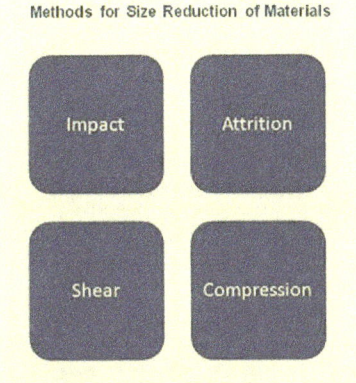

Figure: Methods for size reduction of materials

The reduction of size of the material when it pass through a crusher is expressed as reduction ratio. The reduction ratio is the ratio of the crusher feed size to product size. The sizes are usually defined as the 80 % passing size of the cumulative size distribution.

Types of Crushers

Crushers are classified into three types based upon the stage of crushing they accomplish. These are (i) primary crusher, (ii) secondary crusher, and (iii) tertiary crusher. The primary crusher receives material directly from run of mine (ROM) after blasting and produces the first reduction in size. The output of the primary crusher is fed to a secondary crusher, which further reduces the size of the material. Similarly the output of secondary crusher is fed to the tertiary crusher, which reduces the material size further. Some of the materials may pass through four or more of the crushing stages before it is reduced to the desired size. The degree of crushing is spread over several stages as a means of closely controlling product size and limiting waste material.

Crushers are also classified by their method of mechanically transmitted fracturing energy to the material. Jaw, gyratory and roll crushers work by applying compressive forces while impact crushers such as hammer crusher apply high speed impact force to accomplish fracturing.

There are several types of crushers which are used in various industries. These are given below:

Jaw Crusher

Jaw crusher is used as primary crusher. It uses compressive force for breaking the material. This mechanical pressure is achieved by the two jaws of the crusher. Reduction ratio is usually 6:1. The jaw crusher is consisting of two vertical jaws installed to a V form, where the top of the jaws are further away from each other than the bottom.

One jaw is kept stationary and is called a fixed jaw while the other jaw, called a swing jaw, moves back and forth relative to it, by a cam or pitman mechanism. The volume or cavity between the two jaws is called the crushing chamber. The movement of the swing jaw can be quite small, since complete crushing is not performed in one stroke. The inertia required to crush the material is provided by a weighted flywheel that moves a shaft creating an eccentric motion that causes the closing of the gap.

Feed is entering to crusher from the top and lumps are crushed between jaws. Jaw crushers are heavy duty machines and hence need to be robustly constructed. The outer frame is generally made of cast iron or steel. The jaws themselves are usually constructed from cast steel. They are fitted with replaceable liners which are made of manganese steel, or Ni-hard (a Ni-Cr alloyed cast iron).Usually both jaws are covered with replaceable liners. Also in some types, the liners can be turned upside down after a while, extending the replacement time.

Gyratory Crusher

A gyratory crusher is similar in basic concept to a jaw crusher, consisting of a concave surface and a conical head with both the surfaces are typically lined with manganese steel liners. The inner cone has a slight circular movement, but it does not rotate. The movement is generated by an eccentric arrangement. The crushing action is caused by the closing of the gap between the mantle line (movable) mounted on the central vertical spindle and the concave liners (fixed) mounted on the main frame of the crusher. The gap is opened and closed by an eccentric on the bottom of the spindle that causes the central vertical spindle to gyrate. The vertical spindle is free to rotate around its own axis.

The material travels downward between the two surfaces being progressively crushed until it is small enough to fall out through the gap between the two surfaces. A gyratory crusher is used both for primary or secondary crushing.

Cone Crusher

Cone crusher is consisting of a crushing chamber, a crushing cone and a operating mechanism. The cone is built in to a vertical shaft, which is supported from the top with a bowl-shaped bearing and from the other end to an eccentric operating mechanism.

Cone crusher is similar in operation to a gyratory crusher, with less steepness in the crushing chamber and more of a parallel zone between crushing zones. A cone crusher breaks material by squeezing the material between an eccentrically gyrating spindle, which is covered by a wear resistant mantle, and the enclosing concave hopper, covered by a manganese concave or a bowl liner. As the material enters the top of the cone crusher, it becomes wedged and squeezed between the mantle and the bowl liner or concave. Large pieces of the material are broken once, and then fall to a lower position (because they are now smaller) where they are broken again. This process continues until the pieces are small enough to fall through the narrow opening at the bottom of the crusher.

A cone crusher is suitable for crushing a variety of mid-hard and above mid-hard materials. Feed is dropped to the crusher from the top and it is crushed between the crushing chamber and the slowly rotating cone.

Cone crushers are mostly used for the large scale crushing in the mining industry. It has the advantage of reliable construction, high productivity, easy adjustment and lower operational costs. The spring release system of a cone crusher acts as an overload protection that allows tramp to pass through the crushing chamber without damage to the crusher.

Roller Crusher

Roller crusher is a crusher that breaks material by squeezing it between two revolving metal cylinders, with axes parallel to each other and separated by a space equal to the desired maximum size of the finished product. It consists essentially of two opposite directions driven cylinders that are mounted on horizontal shafts. The other shaft is mounted permanently in the frame and is leaning on robust springs. The gap between cylinders can be adjusted, so the size of crushed product is easily adjustable. Usually both cylinders are covered with manganese steel liners. Crushing ratio is usually lower than in other crushers. Roll crusher is suitable for fine crushing. The roll crusher uses compression to crush materials. Reduction ratio is 2 to 2.5 to 1. Roller crushers are not recommended for abrasive materials.

Hammer Crusher

Hammer crusher consists of a high-speed, usually horizontally shaft rotor turning inside a cylindrical casing. The crusher contains a certain amount of hammers that are pinned to the rotor disk and the hammers are swinging to the edges because of centrifugal force. Feed is dropped to the crusher from the top of the casing and it is crushed between the casing and the hammers. After crushing the material falls through from the opening in the bottom.

Impact Crusher

Impact crushers make use of impact rather than compression to crush material. The material is contained within a cage, with openings of the desired size at the bottom, end, or side to allow crushed material to escape. There are two types of impact crushers namely (i) horizontal shaft impact crusher, and (ii) vertical shaft impact crusher.

Impact crushers are often used with materials, which are soft or which are easily cleaving from the surface. The crusher consists of a fast spinning rotor and beaters attached to the rotor. Feed is entering to the crusher from the top and crushing starts immediately when the feed is impacted with beaters towards the crusher's inner surface. Impact crusher can also be equipped with a bottom screen, which prevents material leaving the crusher until it is fine enough to pass through the screen. This type of crusher is usually used for soft and non abrasive materials.

Mineral Sizers

The basic concept of the mineral sizer is the use of two rotors with large teeth, on small diameter shafts, driven at a low speed by a direct high torque drive system. This design produces three major principles which all interact when breaking materials using sizer technology. The unique principles are the three-stage breaking action, the rotating screen effect, and the deep scroll tooth pattern.

- The three-stage breaking action – Initially, the material is gripped by the leading faces of opposed rotor teeth. These subject the material to multiple point loading, inducing stress into the material to exploit any natural weaknesses.

At the second stage, material is broken in tension by being subjected to a three point loading, applied between the front tooth faces on one rotor, and rear tooth faces on the other rotor. Any lumps of material that still remain oversize, are broken as the rotors chop through the fixed teeth of the breaker bar, thereby achieving a three dimensional controlled product size.

- The rotating screen effect – The interlaced toothed rotor design allows free flowing undersize material to pass through the continuously changing gaps generated by the relatively slow moving shafts.

- The deep scroll tooth pattern – The deep scroll conveys the larger material to one end of the machine and helps to spread the feed across the full length of the rotors. This feature can also be used to reject oversize material from the machine.

Criteria for selection of crusher

The following are the criteria used in the selection of the right type of a crusher for crushing a material:

- Production requirement – It includes output size and shape, and the required capacity.

- Ore characteristics – Include material specification, feed (input) size, material friability, and material abrasiveness.

- Operational considerations – It includes power demand, equipment availability (hours/annum), availability and cost of replaceable parts, reduction ratio, maintenance requirements, needed manpower, approachability of parts for maintenance, availability of spares, safety and environment.

- Equipment ruggedness – Can the crusher pass uncrushable debris without damage to the crusher.

- Capital cost of the crusher and the total cost of the crusher station

Jig Concentrators

Mining jigs make use of gravity recovery in order to separate heavy minerals from the lighter gangue (waste material). Conventionally, mining jigs look similar to an upside down cone that is filled with water. Near the top is a grid, and material is placed above that grid. At the bottom is a motor which creates a pulse throughout the cone, allowing for small waves to be pulsated and the material to be dispersed on each pulse. The lighter particles take longer to settle and tend to hover near the top, whilst the heavier particles settle faster on the grid. Because the material is continuously dispersed, the heavy material is gradually trapped at the base of the grid, and the lighter particles are washed off the end of the jig and put aside as waste material.

Jigs are used to recover a variety of minerals that lie in alluvial deposits. Hard rock can also be ground to recover diamonds, gemstones, gold and coal. Jigging caters for a broad spectrum of minerals from alluvial ores to anthracites to gold and platinum. Jigging is known as the least time consuming method of separation in terms of preparation with most coals and ores of ferrous metals not requiring any form of crushing.

Harz Type Jig

The (Improved Harz Type) Jig is essentially a coarse – mineral Jig Concentrator embodying improvements on the Harz Jig and the Ellis Jig. It will handle successfully all types of ores which can be treated by gravity concentration with feed size ranging from ¼" to 1½".

 These jigs are made in three sizes and with one to six compartments subdivided by a shallow partition into a screen compartment and a plunger compartment. Reciprocation of the plunger causes äow of water through the screen. The coarse and heaviest particles settle onto the screen and are removed by a side drawoff. The heavy ãne particles pass through openings in screen into hopper-shaped hutch, from which they are discharged intermittently or continuously. The light particles pass into next compartment where currents are of lower velocity.

Standard construction is tanks and frames made of kiln-dried Douglas ãr and laminated wood plungers. Iron work consists of necessary tie rods, eccentrics, plunger rods, screens, concentrate draw offs, shafting and bearings. Tanks may be welded steel plate. Jig may have belt or motor drive.

Additional data gladly furnished upon request.

Jig Concentrator Capacity

Size	No. of compartments	* Av. Capac. In Tons per 24 Hours	Screen Area in Sq. Ft.	Motor H.P.	Approx, Ship. Wts., Lbs. Wood or Steel Motor Driven	
					Domestic	Export
18" x 32"	1	40	4	1	1450	1750
	2	80	8	1 ½	2550	3050
	3	120	12	2	3575	4300
	4	160	16	3	4600	5525
	5	200	20	3	5550	6650
	6	400	24	5	6600	7900

24" x 36"	1	60	6	1 ½	2125	2550
	2	120	12	2	3825	4600
	3	180	18	3	5450	6550
	4	240	24	5	6950	8350
	5	300	30	5	8400	10000
	6	360	36	7 ½	9800	11800
30" x 36"	1	75	7 ½	1 ½	2850	3400
	2	150	15	3	5225	6200
	3	225	22 ½	5	7650	9200
	4	300	30	5	9600	11500
	5	375	37 ½	7 ½	11600	13900
	6	450	45	7 ½	13600	16300

*Average capacity is based on 10 tons of ore per square foot of screen surface per 24 hours.

Size	*Dimensions of 1 and 2 Compartment Units					
1 inches	Steel Construction			Wood Constructions		
	L	W	H	L	W	H
	3'8 ½"	3'3 ½"	6'0"	3'4 ½"	5'6"	8'4"
	6'9"	3'3 ½"	6'2"	6'4 ½"	5'6"	8'4"
	4'7 ¼"	4'6"	6'9"	3'8 ½"	6'6"	8'10"
	8'2 ½"	4'6"	6'11"	6'11 ½"	6'6"	8'10"
	4'7 ¼"	5'9 ¾"	7'8"	3'8 ½"	7'6"	9'6"
	8'2 ½"	5'9 ¾"	7'10"	6'11 ½"	7'6"	9'6"

* Dimensions for 1 to 6 compartment units will be gladly furnished on request.

Note: Water requirements vary from approximately 3 to 10 tons per ton ore treated. Stroke variation is from 0"-1" on the 18"x32" size to 0"-1 ½" on the 24"x36" and 30"x36" sizes.

Selective Mineral Jig Concentrators

Mineral Jig is not just another jig. It is a highly efãcient selective pulsator and concentrating machine, which has the ability to treat an unclassiãed feed . . . separating solids having only a slight differential in settling rates, with only a minimum addition of water. It is successfully used on practically every type of ore. Gold, lead, zinc, copper, native copper, nickel, iron, manganese, tungsten, chromite, and äuorspar are a few of the ores which can be concentrated in this jig.

The Mineral Jig is usually installed in the grinding circuit between the ball mill and classiãer, or at any other point in the circuit where free mineral is present. It is ideal for use in cyanide, äotation, and gravity-concentration mills; base-metal and nonmetallic mills; tungsten- gold mills; and placer operation; to save minerals as coarse and as soon as possible, thus reducing grinding cost.

This jig is simple to regulate and control, requiring only periodic discharging of concentrates from the hutch, which can be locked to prevent theft of high-grade mineral, or can be arranged for continuous discharge by using a Dowsett Density Control Valve. The Mineral Jig requires a minimum of attendance; dilution is easily controlled, as it is always necessary to add water to the jig tailing for subsequent classiãcation. It requires minimum äoor space—being less than that for any other concentration unit; and has a low initial cost. No pumps or elevators are needed in the ball-classiãer circuit, as the jig acts as a launder. It treats an unscreened, unclassiãed feed—yielding a clean, high-grade concentrate. The distinctive upper trash screen prevents foreign tramp matter from interfering with normal jig operation, and the lower wedgebar screen minimizes the blinding so common in other jigs. The rugged design of the jig with its few wearing parts insures continuous "24 Hour Service."

Jig Concentrator Capacity

Machine Size	*Capacity With Circulating Load Tons Per 24 Hrs.	Dimensions			** Motor H.P.	Approximate Shipping Weight Pounds		
		L	W	H		Gas	Belt	Motor
8"x12" Simplex	7-35	24"	34"	43"	½	535	430	500
8"x12" Duplex	15-45	35"	36"	53"	¾	850	700	815
12"x18" Simplex	24-75	26"	42"	66"	¾	1165	1015	1130
12"x18" Duplex	50-150	48"	42"	66"	1	1650	1375	1600
16"x24" Simplex	75-200	32"	51"	66"	1	1500	1225	1450
16"x24" Duplex	150-400	60"	51"	66"	1 ½	2000	1790	2030
24"x36" Simplex	200-400	44"	69"	67"	2	2900	2500	2750
24"x36" Duplex	400-800	86"	69"	67"	2	3975	3575	3820

*Capacity should be decreased 50% when jigging non-metallics.

**H.P. of gasoline engine is 1.34 up to and including 16"x24" Simplex.

2 H.P. for 16"x24" Duplex and 3 H.P. for 24"x36" Simplex and Duplex.

Harz Jigs differ in principle from all those previously described, inasmuch as the particles are separated by their fall through a somewhat deeper column of water than is the case on inclined tables, while a series of blows from below, causing waves moving upwards, continually brings the particles into suspension, and allows them to drop again. The initial period of the fall in water, during which the motion depends chieäy on density, is thus continually reproduced, and the result is a perfect separation of heavy from light particles of ore when working on any materials except the änest pulp. Jigs consist of sieves supporting beds of ore, which are completely immersed in water ; the ore is raised and allowed to fall by a quick succession of currents of water caused by the sudden action of a piston below, which is so worked that the upward movement or pulsation resembles that produced by a blow, while the downward movement is gradual. Under these conditions the heavy particles work downwards and pass through the sieve, while the lighter gangue is carried away horizontally by a stream of water introduced either from below or from above. Such machines are especially suitable for coarse ores.

In the Harz jig a layer of coarse heavy particles is spread on the sieve to prevent too much of the ore from passing through. The stuff is fed in regularly at the head of the jig, and the strokes of the piston raise both the bed of heavy particles and the ore. The heaviest grains of ore änd their way during the downstroke into the interstices of the bed, gradually pass through it, and coming to the screen, fall through into the tank below. The lighter particles cannot descend, and are gradually washed over the end partition by the continuous supply of water. Two products are therefore given, neither requiring further treatment if the conditions are favourable, and the machine properly adjusted. The wire meshes of the screen are always much larger than the ore treated, and the bed is composed of material of as nearly as possible the same density as the concentrates to be obtained, and is usually from ½ to 1 inch in depth. The number of strokes of the piston per minute is from 60 to 80 with coarse sand of 1/12 inch in diameter, and 200, 300, or even 400 with very äne sand, approaching slimes. The length of stroke varies under the same conditions from 2/5 to 1/5 inch, and in the case of very äne, almost impalpable sand, the stroke may be diminished till it becomes a mere tremor. Some of the highest authorities on concentration have stated their belief that for enriching even very äne sand, the Hartz jig is the simplest and most economical machine yet invented, and requires the least amount of labour.

Selective 4″x 6″ Mineral Jig Concentrators

Mineral Jig, size 4″ x 6″, is a unit which älls the needs between the 8″x 12″ Simplex Mineral Jig and the No. 1-M Laboratory Jig, due to its applicability to both commercial and test work.

Although the 4″ x 6″ Mineral Jig is ideal for many uses in commercial work, it also has considerable merit as a unit for use in pilot mill operation and in large batch laboratory tests. It is suited for non-metallic or open-circuit work in the laboratory especially where a large quantity of concentrate must be obtained. This size jig is most desirable for trial installation in large tonnage mills to determine the desirability of a jig in the circuit, as it can operate on small quantities of material of the same size as that handled by the larger size jigs.

The 4″x 6″ jig has a single hutch compartment and a reversible screen compartment constructed of cast iron so that cyanide solutions can be fed to the unit instead of water if desired. The rotating water valve is timed to admit water only on the plunger up-stroke and is located on the end of the eccentric shaft. The eccentric shaft is mounted in a ball-bearing double-pillow-block type bearing. An adjustable eccentric is used for varying the movement of the rubber diaphragm. The upper trash screen consists of a four mesh woven-wire cloth. Two lower wedge-bar screens are furnished with each jig; one with 2-millimeter openings and the other with 5-millimeter openings.

When the jig is used on continuous tests, a steel stand is recommended for mounting the jig. The steel stand is arranged with a shelf for holding a glass jar which accumulates the jig concentrate as it discharges from the hutch and which facilitates the inspection of the type of jig product while the test is being conducted.

Jig Capacity

Capacity Pounds Per Hr.	Steel Stand	Dimensions			Motor H.P.	Shipping Wt. Lbs.	
		L	W	H		Belt	Motor
150-500	With	18"	26"	46"	¼	290	315
150-500	Without	18"	23"	24"	1/4	175	200

Richards Type Pulsating Concentrating Jig The (Richards Type) Pulsating Jig concentrates ore too coarse for treatment on concentrating tables. It is similar to the ordinary plunger type of jig except that the plunger action is supplanted by a pulsating water current. By this means the screen surface has a much greater capacity than that of the ordinary jig. As there is no suction stroke, and the pulsating current is always upward, there is no clogging of the screen. A high grade mineral product is made which is drawn from the hutch. This jig is made in various sizes. Its water consumption is greater than the (Selective) Mineral Jig.

References

- Drill-and-blast-method: railsystem.net, Retrieved 13 May 2018

- Types-of-mining-equipment: editorstop.com, Retrieved 07 March 2018

- Drilling-rig-1278: petropedia.com, Retrieved 19 May 2018

- Types-drilling-rigs-structures-ali-seyedalangi: linkedin.com, Retrieved 21 May 2018

- What-are-bucket-wheel-excavators-and-how-do-they-work: natparts.com, Retrieved 17 July 2018

- Crushers-and-their-types: ispatguru.com, Retrieved 17 June 2018

- Mining-equipment-terminology: aptprocessing.com, Retrieved 10 April 2018

Chapter 4

Transportation Systems in Mines

In order to completely understand mining and mining processes, it is necessary to understand the mechanical infrastructure intrinsic to transportation systems in mines. Such transportation systems include conveyor system, trams, headframe, loader, winding engine, etc. The following chapter elucidates these tools and machinery associated with transportation in mines.

Conveyor System

A conveying system is an automated system of conveying something from one area to another. It utilizes mechanical energy, often via a system of belts and pulleys, thereby avoiding the necessity of human or animal labor while simultaneously achieving highly predictable, repeatable speeds and performance levels. Conveying systems may be designed to transport solids or liquids and may move massive materials, lightweight materials, or anything in between. The speed and capacity of the conveying system can be adjusted depending on the needs of the circumstance and the conveying system may operate horizontally, vertically, or at an incline.

Since introduced in the early 20th century, conveyors have become a vital part of mine operations. Conveyor systems can be used to transport material in underground and open pit operations, and are especially common in coal mines (Betz, 1998). The simplicity and increased reliability of conveyor systems make it an attractive alternative to conventional truck haulage. While the capital cost associated with installing conveyor systems are similar to those of haul trucks, significant savings can be achieved from lower operating costs.

Types of Conveyor System

Conventional Mining Conveyor

The conventional conveyor is the most common conveyor system, uses two or more pulley systems with a belt that rotates about them, carrying medium. The two pulleys can be powered for certain requirements in the conveyor system. The conventional conveyor belt is 'v' shaped to better hold the material. Currently, most long conveyors use steel-cord belts. Sorter conveyors often use polyester/nylon fabric known as EP fabric. Different variations of the belt conveyor have been conceived to deal with challenges unique to certain environments. These variations common to the mining industry include the gravity conveyor, cable belt, pipe conveyor, and sandwich conveyor/apron conveyor.

Conventional Mining Conveyor System

Gravity Conveyor

Similar to the conventional mining conveyor, the gravity conveyor is a two pulley system. However, gravity conveyors travel on a decline benefiting from potential energy. This belt turns a motor within a generator storing electricity as ore is transported. Depending on the height and steepness of the belt slope, the value of the energy produced is often greater than the operating cost of the conveyor. This allows the gravity conveyor system to operate at a profit.

Cable Belt

The cable belt is a conveyor system where the belt is supported between two steel cables on either side. In this variation, the cables absorbs the driving forces of the belt. These are not as common as the conventional steel-cord belts, as capital and operating costs are higher. These conveyor systems can be covered to reduce outside contamination as shown in the picture to the left.

Cable belt Conveyor System

Pipe Conveyor

In the pipe conveyor, the belt is converted into a cylindrical shape in order to fully enclose the material being transported. Benefits of this are the following:

- Completely eliminates spillage and is more environmentally friendly.

- Allows for sharper vertical and horizontal bends because there is no risk of spillage.

- Can take a more direct path when navigating irregular terrain reducing civil works.

- Can climb slopes 50% greater than trough belts.

- The belt can be looped and the underside can be set up to transport material in the opposite direction.

Pipe Conveyor System

Sandwich Conveyor

The sandwich conveyor is useful when navigating steep slopes. A second belt rides on top of the material in steep sections to hold material in place. This is often assisted by what's called inverted pressing idlers.

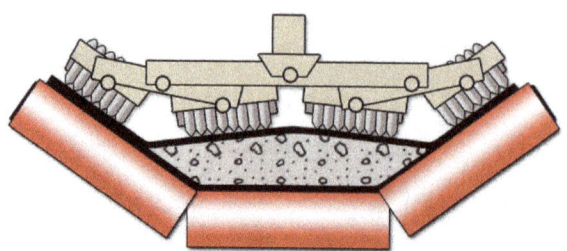

Cross Section of a Sandwich Conveyor System

Components in a Conveyor System

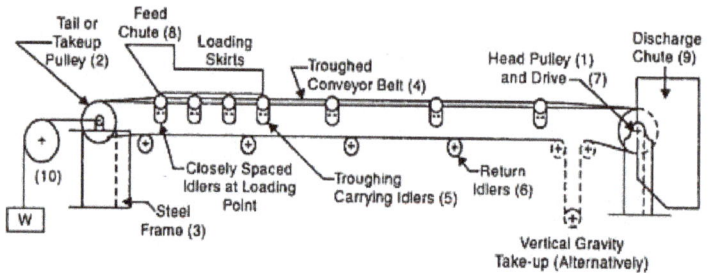

A standard conveyor system in the mining industry

Head and Tail Pulley

There are several components that makeup a conveyor system. The head pulley, tail pulley and belt are the essential components of a conveyor system. The head pulley, located at the discharge point, powers the rotation of the belt and is usually fixed. The tail pulley, located at the feed chute is movable. However, the tail pulley is usually held in place for a long period of time. The tail pulley is fixed by a counter weight and pulley system, or in situations where space is limited (underground operations) by a piston or winch. Using this variable system, the position of the tail pulley is adjusted to ensure tension and slack of the belt compensate for changing conditions.

Head Pulley

Belt

The conveyor belt carries material between the head pulley and tail pulley. The conveyor belt is composed of two layers of material; the under layer and the over layer. The under layer provides linear strength and shape, it is composed of woven fabrics that have warp and weft such as polyester, nylon or cotton. The over layer, often referred to as the 'cover', is composed of a material such as rubber or silicone (lower frictional surfaces) and heat/gum rubber (higher frictional surface).

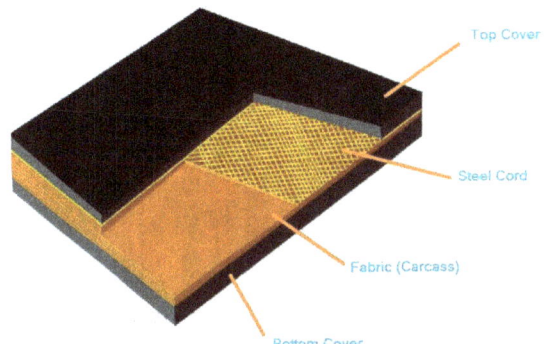

Cross Section of Conveyor Belt

Idlers

Idlers support the conveyor belt. Additionally, idlers control the slack in the belt. In Lehman's terms, idlers are rollers that spin using ball bearings. They are strategically spaced based on the forces that are expected to occur at different points in the belt. Spacing of idlers will vary depending on the forces impacted on the belt at a given place. Areas of varying spacing include the carrying zones, impact zones and spacing for return idlers. Furthermore, different types idlers of idlers are used depending on the required belt trough angles.

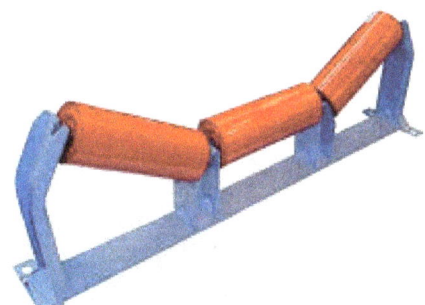

Example of a standard trough idler

Belt Cleaner

Belt cleaners (otherwise known as belt streamers) remove any carry back material. Carry back material, muck that sticks to the belt at the discharge point, is returned to the head pulley through the underside of the belt. As such, belt cleaners are typically located underneath the head pulley. Carry back is detrimental for a variety of reasons:

- It causes excessive buildup on the belt and pulleys and can cause blockages.

- Causes belt misalignment due to an artificial crown created by the carry back.

- Accumulation of material under the conveyor.

- Overall reduced operating efficiency and profitability from higher maintenance costs and lost material.

Belt Cleaner

Discharge Chute

The discharge chute creates a smooth transition of material from either one conveyor system to a stockpile or one conveyor system to another conveyor system. It is located under the head pulley at the discharge point. The discharge chute is most important in conveyor to conveyor transport as it reduces impact on the second conveyor system (increasing wear on the belt and reducing maintenance cost). Furthermore, poor transfer design can lead to premature belt wear, poor ore tracking, material degradation, and dust generation.

Discharge Chute

Feed Chute

The feed chute, located near the tail pulley, brings material onto the conveyor system. Much like the discharge chute, it is a very important component of the conveyor system. As transfer design is critical to the system, installing a feed chute is imperative to reducing reliability costs.

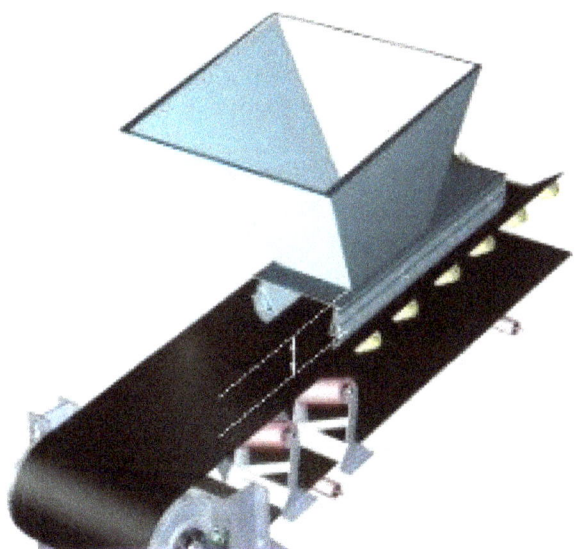

Feed Chute

Safety Systems

Safety systems are implemented to minimize hazards around conveyor systems. Some innovations include emergency stop cords and hazardous zone barricades (or screens). At the beginning of some conveyor systems, a magnetic sensor is used to detect if any rock bolts or other metallic materials are on the belt. Since these materials can rip the belt, if detected, they are immediately expelled through a secondary chute.

Design of a Conveyor System

The Goodman Conveyor Company, an industry expert, has outlined the steps in designing a conveyor system. A simplified version has been summarized below, along with contributions from others sourced below:

1. Determine the desired conveyor capacity.

2. Identify the material and its characteristics.

3. Choose a troughing angle.

4. Determine belt width.

5. Select belt speed.

6. Determine the idler spacing.

Conveyor Capacity

The first step in designing the conveyor's capacity is to determine the desired capacity required to transport the desired amount of material. This capacity, in t/h, should be the peak surge volume that is expected. Conveyor capacities can range anywhere up to 40, 000 tonnes per hour for a 3200 mm wide and 45 mm thick belt. By finding conveyor capacity, sizing of other conveyor parameters can be calculated.

An important design consideration for a conveyor system is to characterize the material handled. Furthermore, parameters such as the over cover belt material, the maximum incline a conveyor can scale, and trough design can be identified from these material characteristics. For example, a abrasive material such as ore from an in pit conveyor system will have much different characteristics than coal. Characteristics of the material that must be considered include:

- Angle of repose: static angle between the free formed pile of the material and the horizontal.

- Angle of surcharge: dynamic angle of repose (usually 5 – 10 degrees less than angle of repose).

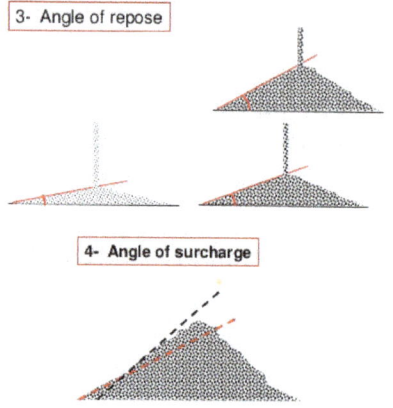

Angle of Repose and Angle of Surcharge

Trough Angle

The trough angle is the angle of incline on either side of the belt (as mentioned earlier conventional belts forms a 'V' shape). Trough angles between 35 and 45 degrees generate a higher carrying capacity than a flat belt with equal widths. However, belts with higher trough angles require belt material that can provide a higher flexibility. An optimal trough angle is calculated based on material characteristics and limitations of the material chosen for the belt.

Belt Width

Belt width is a function of the largest lump size (largest product particle size), operating speed, and the desired belt capacity. The largest belt belt width of all conveyor system is 3.2 metres. A general rule of thumb for calculating belt width is that the belt should between three to five times the height of the maximum lump. If there are more particles of larger sizes in the mine operation, a higher width should be used.

Belt Speed

Typical speeds for conveyors can vary from one meter per second (small conveyors) to 8.5 metres per second (overland conveyors). Spreading conveyors, can reach speeds of up to 15 metres per second with a handling capacity of up to 40,000 t/h.

Belt speed is determined by several different factors and benefits. Increasing belt speed, increases belt capacity, which decreases belt width and tension. Moreover, this increases the importance of the design of transfer points, and generally reduces the life of all conveyor components. As such it is important to find a balance between, capacity required and reliability costs.

Idler Spacing

While spacing can vary depending on the load requirements, many belt systems will use four feet spacing for carrying zones, one foot for impact zones (such as loading) and eight to ten feet for return idlers on the underside of the belt. Furthermore, idler spacing is determined based on belt weight, material weight, idler rating, belt sag, idler life, belt rating and belt tension. Moreover, a general rule of thumb is to limit belt sag to 2% of idler spacing.

Technical Design of a Conveyor System

In order to ensure safe operation and retain the highest reliability ratings in the system, it is imperative that conveyors are designed to handle all types of loads and forces. In the following section, equations are introduced to determine; forces on the conveyor belt, the minimum required drum diameter, drum power required, and motor sizing.

Belt Tension

Any time an object is in tension, it is vital to find the forces acting on the item and and the energy the item possesses. The forces felt by the conveyor belt are a function of the skid plat, the friction of the idler rollers, the mass of the belt, the mass of the rotating drum roller, and the mass of the materials being conveyed over the belt (Siegling Belting - Transilion). All of this is used to determine the effective pull in the equation below:

$$F_U = u_T * g(m + m_B / 2) + u_R * g(m_B / 2 + m_R)$$

Where:

- u_T is the coefficient of friction of the skid plat.

- u_R is the coefficient of friction of the support rollers.

- m_B is the mass of belt.

- m_R is the mass of all rotating drum/rollers not including the drive drum.

- m is the mass of the material transported by the belt at a given time.

- F_u is the effective pull.

This is then used to calculate the maximum belt force using the equation:

$$F_1 = F_U * C_1$$

Where C1 is a calculation factor provided by the belt manufacturer.

Next, a second calculation factor, C2, is calculated and compared to C1 using the following equation:

$$C_2 = F_1 / b_o$$

Where bo is the width of the belt.

If C2 is below the proposed belt constant, then the proposed belt can withstand the stresses felt in the conveyor system.

Horizontal Bends

Horizontal bend conveyors are used to bypass obstacles and reduce the number of broken routes in the conveyor system. However, horizontal bend conveyors depend on a number of factors which include; belt stretching forces (elastic properties of the belt), horizontal curve radius and the distance between support rollers. All these factors are transitively characterized in the equation below. As horizontal bend conveyors are expensive, it is necessary to justify the value of implementation.

$$F_x = T * R_c / \left(R_c^2 * (x - L_c)^2 \right)$$

Where:

- T is the actual distance of belt at the distance x. N.

- L_c is curve length (in metres).

- R_c is the curve length (in metres).

- x is the distance in metres of the observed point from the beginning of the curve (in metres).

- F_x is the value of the force acting in some point in the curve.

It is important to note that for horizontal bends the maximum belt force should be higher then a straight conveyor system. This should be verified by finding both the C_1 and C_2 values for this special circumstance and comparing the values.

Drum Diameter

A correctly sized drum diameter is used to calculate the minimum power required for the system to operate. The drum diameter is a function of belt width, the effective pull of the belt, and the arc of contact on the drum (Siegling Belting - Transilion). The drum diameter is determined using the following equation:

$$d_A = (F_U * C_3 * 180) / (b_o * b)$$

Where: C_3 is a calculation factor provided by the manufacturer b is the arc of contact on the drum

Drum Drive Power and Motor Size

Parameters used to determine the drum drive power include the drum diameter calculated in the above section, and the belt speed (Siegling Belting - Transilion) It is calculated using the following formula:

$$P_A = (F_U * v) / 1000$$

Where v is the belt speed.

The drum drive power is used to find size the motor required. Assuming an efficiency of 80%, the motor size is calculated using (Siegling Belting - Transilion):

$$P_M = P_A / 0.8$$

It is imperative to create a conservative motor design. When calculating motor size, one should always round to the next largest motor. A design calculation can be seen to the right. Although the design is crude, it is a good estimation for compromise between excessive capital expenditure and system reliability. Under designing a conveyor motor size can result in inadequate belt speeds and failure. While over design can result in voltage dip problems.

Maintenance

Maintenance – Step by Step Guide

Maintenance on conveyor belts can be characterized into three broad steps.

The first step, the most complicated one, involves shutting down the conveyor system and emptying the medium on the belt. After the belt is shut down; bearings, universal joints and pulleys are checked and lubricated.

Next, the tension, wear and lubrication in the belt is checked. Moreover, sprocket alignment, wear and screw set is then checked. If a v-belt is within the system; the tension, wear and sheave alignment is checked. Electrical connections to the conveyor are checked. The gearbox is checked and oil is added to the proper level. Lastly, the general condition of the system is checked.

The third step is starting the system and ensuring it runs properly; if this is not the case then the system needs to be turned off and checked for any irregularities (i.e. back to step one). The last step involves adding medium and continuing with production.

Types of Maintenance

There are 4 different types of maintenance, each summarized below:

- Preventative maintenance

- Opportunity based maintenance

- Corrective maintenance

- Condition based maintenance

Preventative Maintenance

Preventative maintenance is calendar based maintenance. Part replacements occur after certain run times. This type of maintenance relies solely on reliability data provided by the manufacturer. The manufacturer's preventative maintenance run times are generally conservative. This means many parts are being replaced well before they are at a serious risk of failing. While some parts of the conveyor system can be replaced while the conveyor system is operational, preventative maintenance often requires shutting down the system. This results in a costly loss of production. However, when compared to run until failure, part replacement intervals are lower and overall failure time is lower.

Opportunity Based Maintenance

Opportunity based maintenance refers to taking advantage of overall system down time to replace parts. An example of this would be if mining operations halted, the conveyor system would be shut down and thus replacements could be made to the conveyor system. Similar to preventative maintenance, the replacement of any part is not strictly based on run time. As such, parts may not need to be replaced. Furthermore, this will result in greater up time, but runs a higher risk of costs associated with part failure.

Corrective Maintenance

Corrective maintenance (otherwise known as run to failure maintenance) is when the system is run until components within the system undergo failure. When failure occurs, downtime is generally much longer and more expensive than the previously discussed maintenance types. If failure is not detected immediately, catastrophic damage can be inflected on the system which can amplify conveyor downtime and costs.

Condition Based Maintenance

Condition based maintenance involves monitoring different components of a system to predict when failure might occur. When failure can be predicted, the appropriate corrective actions can be taken to minimize the effect on production. Recent studies show that up to 90% of mechanical failure can be predicted. In many cases, maintenance can be carried out without interrupting production. Condition based monitoring uses heat and vibration sensors to predict when a component is at risk of failure. Vibration signatures received from the sensors in an area of the conveyor will be much different when the conveyor is running smoothly compared to when it requires maintenance. This would give accurate close to failure readings of major components such as major drives, idlers and pulleys. Heat sensors are used, as parts will often fail when they reach a certain temperature. By monitoring the temperature of vital parts, maintenance can be carried out before parts reach a critical temperature. Condition based maintenance interrupts production for maintenance only when it is necessary.

Factors Effecting Reliability

Although, conveyor systems provide reduced downtime when compared to haul trucks, it is imperative to limit downtime on these systems. Reliable and available equipment is essential in conveyor systems because they deliver small amounts of material over long periods of time (Overland Conveyor). Furthermore, choosing a suitable type of maintenance for the mining operation with reduce operational costs while maximizing uptime. Unlike haul trucks, production halts when one conveyor fails. This is why there is a great focus on reliability when it comes to conveyor systems.

Maintenance Type

The type of maintenance scheduling has a great effect on the overall reliability and availability of the conveyor system. Corrective maintenance results in the worst reliability and availability in a mining conveyor system. As failure is unexpected, maintenance crews are not as prepared when compared to preventative maintenance. Moreover, there is also a significantly higher diagnostic time associated with run to failure maintenance. Lastly, corrective maintenance increases the risk of a domino effect of secondary failures caused by the original failure.

Condition based monitoring, in theory, provides the greatest reliability and availability. Being able to accurately predict when a component is reaching failure based on sensors is objectively better than estimating when it will fail based on past experience. In practice, the accuracy of the predictions is a function of the accuracy of the sensors used. In the past, trouble collecting quality and meaningful sensor data has made it difficult to justify the additional expense and effort associated with condition based maintenance. However as sensor technology has improved significantly, condition based maintenance now provides the greatest reliability and is the most economical maintenance option (Schools, 2015). According to the Conveyor Equipment Manufacturers Association (CMEA), well maintained conveyor systems can operate reliably at 90% availability.

Number of Conveyors in Parallel

In a system made up of numerous elements, the reliability of the system depends on the number of elements in the system and the reliability of each element. Assuming each conveyor has the same reliability, a conveyor in a parallel system can be assigned a coefficient of readiness, ri. The reliability of a number of conveyors in parallel is then modeled with:

$$r = r_i^n$$

Therefore, it can be seen that the reliability of the system decreases with each additional conveyor added. To maximize reliability, it is advantageous to make each conveyor as long as practically possible to minimize the number of components in the system.

Common Electrical and Mechanical Issues

There are several issues associated with interplay between electrical and mechanical parts contingent towards system reliability. The most common issues are listed in the table below.

Table: Electrical and mechanical Issues associated with conveyor systems

Electrical Issues	Mechanical Issues
Voltage Dip from High Currents	Gearbox variations
Stalling due to externally imposed voltage fluctuations	Sprocket variations
Difference in voltage supplied by two power pulley systems	Fluid coupling variations
Variations in motors characteristics	Wet clutch couplings
Large transient pulsation in the systems	
Thermal effects	

Of the electrical and mechanical issues listed, the bolded problems are the most common. The voltage dip from high currents is caused by severe starting problems from conventionally designed conveyor systems. Gearbox variations is caused by starting problems or undue stress on the dominant motor and drive train. The slippage of wet clutch coupling generate new difficulties in such operations. All these issues have an effect on system reliability. The underlying issue is caused by having multiple components in the conveyor system, thus triggering electrical and mechanical trips.

Variable Speed Drives

A variable speed drive is a piece of equipment that controls the speed of machinery. These controllers have a combination of computer and power electronic technology that have advanced control algorithms. In the past, they were known to have poor reliability. However, with recent technological advances, they are known to reduce downtime in conveyor systems and actually improve overall reliability. Overall, the advantages of variable speed drives surpass the disadvantages making it more economical to use variable speed drives.

Reliability of Horizontal Conveyor System

With increasing remote mining locations and increasing depths of underground mines, processing plants are increasing in depths from mining operations. This results in material to travel long distances for processing. Horizontal conveyors allow conveyor system to take a more direct route and navigate terrain more efficiently. Thus, this reduces the number of components required in the system, increasing reliability of the overall conveyor system.

Belt Conveyor vs Haul Trucks

Efficiency and costs between conveyor systems and haul trucks are still disputed between professionals in the industry. The disadvantage of conveyors is that they are not as flexible, and require a greater capital cost. In surface operations, installation of in pit crushing and conveying systems may cost more initially than the capital cost of trucks.

However, in pit crushing systems operate more efficiently and at a reduced cost. Trucks are less efficient for the following reasons:

- Only 40% of the energy consumed by trucks is expended during hauling payload. The other 60% is spent hauling the truck body.

- Trucks tend to be empty on return.

- 80% of energy consumed by conveyors is used by delivering payloads.

- Energy costs for trucks are 3 times greater on flat land and 8 times greater on pit slopes than energy costs for conveyors.

The cost benefit of conveyors is generally seen as tonnages increase and the haul distance to the plant increases. The McIntosh Engineering Hard Rock Miners Handbook provides several rules of thumb for when it is advantageous to use conveyors over haul trucks:

- When underground daily mine production exceeds 5000 tonnes.

- When conveying distance exceeds 2 km.

- Beyond 1 km, the cost of transporting a tonne of material by conveyor is 1/10th the cost of transporting it by truck.

Conveyor System Costs

Capital Costs

In-Pit Conveyor:

Capital costs of increased conveyor system increase as tonnages increase. Costs are summarized in the table below of an In-Pit conveyor system with fixed conveyors using a steel idler system and a base length of 610 metres.

Table: Fixed Conveyor System with Rigid Steel Idlers for Material Weighing 801 kg/cu m

Belt Width (cm)	Capacity (mtph)	Motor Size (HP)	Capital Cost ($)	Additional Cost ($/m)
76.2	454	150	1,329,000	1,046
91.4	680	200	1,425,000	1,128
107	907	250	1,544,000	1,177
122	1,361	300	1,728,000	1,297
137	1,814	400	2,282,000	1,827
152	2268	500	2,533,000	1,932
183	2,722	600	2,815,000	2,100
183	3,629	700	2,957,000	2,217

Different types of In-Pit conveyor system will have varying costs, a shiftable In-Pit conveyor systems will have different costs than a fixed In-Pit conveyor system. Costs are summarized in the table below of an In-Pit conveyor system with shiftable conveyors using a steel idler system and a base length of 610 metres.

Table: Shiftable Conveyor System with Rigid Steel Idler for Material Weighing 1,602 kg/cu m.

Belt Width (cm)	Capacity (mtph)	Motor Size (HP)	Capital Cost ($)	Additional Cost ($/m)
61	454	150	1,559,000	1,289
61	680	150	1,559,000	1,289
76.2	907	200	1,712,000	1,378
91.4	1,361	300	1,896,000	1,610
107	1,814	350	2,061,000	1,699
107	2,268	400	2,119,000	1,699
122	2,722	500	2,312,000	1,765
137	3,629	700	2,920,000	2,371

It is important to note that varying the material used to construct the conveyor system will also vary the capital cost of the conveyor system.

Overland Conveyors:

Similar to capital costs of associated with In-Pit conveyor system, overland conveyor system increase in capital costs as tonnages increase. Two tables comparing the capital costs associated with overland conveyors, both with a base length of 1,615 metres, and materials weighing 801 kg/cu m and 1602 kg/cu m, respectively.

Table: Overland Conveyor System with Material Weighing 601 kg/cu m

Belt Width (cm)	Capacity (mtph)	Motor Size (HP)	Capital Cost ($)	Additional Cost ($/m)
76.2	454	500	3,159,000	1,279
91.4	680	600	3,340,000	1,296
107	907	700	3,505,000	1,460
122	1,361	900	3,778,000	1,515
137	1,814	900	4,382,000	1,515
152	2,268	1,400	5,157,000	2,198
183	2,722	1,400	5,470,000	2,171
183	3,629	1,800	6,142,000	2,499

Table: Overland Conveyor System with Material Weighing 1601 kg/cu m

Belt Width (cm)	Capacity (mtph)	Motor Size (HP)	Capital Cost ($)	Additional Cost ($/m)
61	454	500	2,934,000	1,174
61	680	500	2,934,000	1,174
76.2	907	600	3,399,000	1,460
76.2	1,814	900	3,700,000	1,512
91.4	1,814	900	3,700,000	1,512
107	2,268	1,200	4,111,000	1,647
107	2,722	1,200	4,184,000	1,647
122	3,629	1,600	4,916,000	2,355

When comparing the two tables, it is important to note, that when material weights double there is an increase in capital costs as well as additional costs per metre for implementing a conveyor system.

Trams

Many different types of Transportation systems are used in a mine. One of the most used pieces of system in some mines are the mine trams. A tram is basically a four or six wheeled cart that runs on a rail track, much like a train car, except much smaller. Many trams are now operated that are powered by either compressed air or electric carts.

The use of trams has revolutionized many different mines, because they are able to transport not only coal or mined material, but also workers. This is important when

workers are injured they can be swiftly removed from the mine and this has helped many mines achieve a positive safety record.

Trams are mostly small wheeled carts that are used in mines, to transport people, material, coal and mined substances along a track. When trams are used as part of a mines machinery it is often on an electrical system, that powers either an air compressor that pushes the lead "locomotive" on the tram, or an electric motor on the tram lead car.

Trams are thought to be a modern innovation that has helped make mines a lot more safe. Where trams are used the safety record is often a lot better than where they are not used.

Trams are one of the different types of mining innovations that have helped make mines a safer place to work. Workers can leave the mine and return more swiftly allowing breaks and meals to be taken on the surface, and to evacuate injuries if and when miners are in need of medical help. This can also serve to improve the efficiency of the mine as materials are able to be removed swiftly and efficiently, lessening the burden on the miners themselves.

alamy stock photo

Headframe

The main objects of a head frame, or poppet head, are to support the winding pulley firmly and to guide the cage above the surface to the discharging stage. Points, which should be attended to, are:

1. That the pulleys are rigidly supported; the frame being amply strong and rigid for ordinary winding.

2. That the frame shall be strong enough to survive any accidents, which may occur.

3. That the frame is stable against overturning under the worst conditions as to wind or loading.

4. The design must be adapted to local circumstances.

5. It must be durable.

6. The cost shall be as low as is consistent with fulfilling other conditions.

Oage Framing and Pulley Supports

The head frame consists of two Portions: the frame proper, which supports the pulleys, which must be designed to resist all stresses induced by winding; and the cage framing, which really forms a continuation of the shaft above ground, and serves to guide the car above the ground. The headgear may be designed either with the pulley supports independent of the cage framing, or the two may be combined.

If the separate system were adopted, the cage framing must be made strong enough to resist stresses due to supporting the weight of the loaded cage when resting on the keeps and the weight of guide ropes-if such be used. It may also have to bear the shock due to the cage, when overwound and being detached, falling back and being suddenly arrested by the automatic gear.

In the combined system, the pulley supports must be designed to resist stresses due to winding and those above-mentioned.

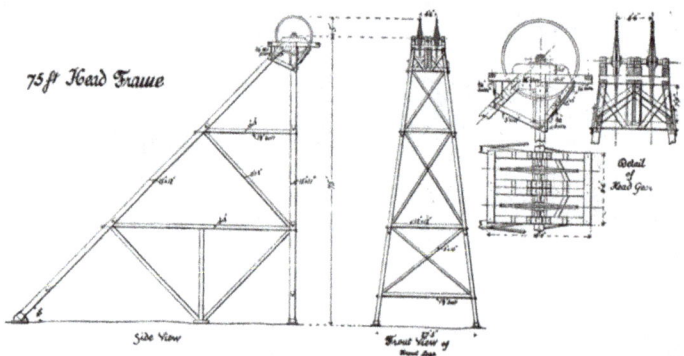

The separate system admits of the cage framing being more compactly designed to fit the cage than is the case with the combined method. The framing is vertical, and is supported on heavy sole pieces placed round the top of the shaft. In this system, the stresses are, more definite and more readily ascertained.

Height

This is one of the first points to be settled, and will depend chiefly upon the situation and particular requirements. It will be necessary to raise the ore above the top of the mill ore-bins, and the height necessary will, of course, be less if there be a slope from the shaft to the mill. There should also be room ' for dumping waste material.

When the height of the discharging level above the surface has been settled, other considerations will decide the height above that level. This height must be sufficient to allow of a certain amount of overwind before the detaching hooks come into operation.

In the best practice with quick winding, it is usually recommended that this be made equal to one revolution of the winding drum. The pulleys should be 'fixed sufficiently above the detaching hook bell platform to allow the cage to be lifted out of the catches without the rope capping coming on to the Sheave of the Pulley.

The height will then be made up as follows: Height above surface to discharge level plus the height of cage and attachments plus the circumference of winding drum plus the distance necessary from detaching platform to center of pulley.

Having determined the height, and whether the frame shall be on the separate or combined system; the members necessary, and their best arrangement must be considered. To do this satisfactorily, a thorough investigation of the forces, which will or may, come upon the structure, is necessary. Assuming that the separate system is adopted, and considering only the pulley supports, there are usually two front legs, vertical in side elevation, and two back-stays sloping hark towards the winder to prevent the front legs being pulled over backwards.

The best position of these backstays is a matter upon which opinion differs. Some designers arrange them parallel to the sloping portion of the winding-rope; others bisecting the angle between vertical and sloping ropes; and others, the majority, in a position intermediate between those two. In some text-books the question is disposed of by giving a rule that the position is determined as follows:

The pull on the vertical and sloping ropes is the same:

Therefore, if we set off distances dc equal to de to equal the pull on the rope, then, df is the resultant, and, theoretically, the position of the back-stay is given by dg. However, to provide against contingencies such as the over-winding of the cage, set out dh equal to 2 dc and take dg' the direction of the resultant under these conditions, as the position of the back-stay. Theoretically, for ordinary working, only one support is necessary, arranged so as to bisect the angle between the vertical and sloping ropes, as is adopted usually for whips. This would be unstable in a large structure, and so a vertical support becomes necessary. If the foot of the back-stay lies between the theoretical position (i.e., bisecting the angle between the ropes) and the shaft, the resultant of the tensions lies outside the base, and there is a tendency for it to be overturned backwards. If it lies between the theoretical position and the winding drum, some of the weight is thrown on the vertical support or front legs.

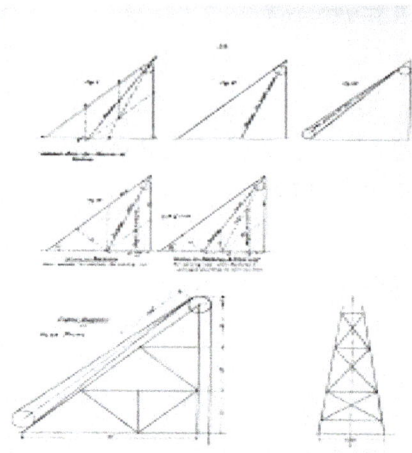

If the directions of the resultant were invariable and of known amount, there is no doubt that the best position for the backstay would be along the line of the resultant. It is possible, however, that accidents may alter its position.

Loading

The principal forces will be the tensions in the winding ropes. If there be two pulleys with one rope· passing under and the other over ·the drum, the greater tension will be in the hoisting rope, the lowering rope being comparatively slack. The mean position of the pull of the sloping ropes, since each rope is alternately hoisting and lowering, is along the line through the center of the winding tangent to the pulley (figure) there will also be the vertical pull down the shaft. The maximum pull will be the sum of the tensions in the two ropes.

For ordinary working, the tension in the ropes will be due to the dead weight of loaded cages, chains and rope, friction against guides and in pulley-bearings, and the force necessary to accelerate the cage from rest to the maximum speed of winding in a certain interval of time. These are the legitimate working loads. Momentary stresses may, however, much exceed these; the commonest being the case where the rope is slack when winding begins. With careless winding the result will be a jerk, producing stresses whose magnitude it is difficult to estimate, but which might easily double the ordinary stress.

Of these loads, the weight of cage, rope, etc., may be considered dead load; that due to acceleration (and retardation) as live load. The shock due to slack rope, besides producing higher stresses, would be more destructive on account of the fact well known that loads suddenly applied have a worse effect than those applied steadily. Hence, an allowance should be made for dynamic effect. To the above loads must be added the weight of the structure itself.

It is necessary to consider, however, not only the commonly occurring stresses, but also the worst possible case.

The greatest possible tension, which may be developed, is equal to the breaking strength of the rope. This tension might be induced by some such accident as the cage suddenly sticking in the shaft owing to some obstruction when travelling at a high speed, with the result that the hoisting rope is snapped.

The engine would itself be incapable of exerting a steady pull equal to this amount, but engine pull plus stresses produced on account of the inertia of the parts, rope, pulleys, etc., might reach this value. The total greatest possible tension on the vertical and sloping ropes may then be taken as equal to the breaking strength of one rope plus the tension on the lowering rope. To this must be added a certain percentage to allow for dynamic effect, say, 25 per cent. Another possible accident is when the cage is overwound and the detaching hook fails to act. The resultant pull would then be practically along the line of the sloping rope, and its limiting value will be equal to the breaking strength of the hoisting rope plus the tension on the lowering rope. The only pull down the shaft would be the tension on the lowering rope. This last case would be that having the greatest tendency to overturn the frame.

Wind pressure must also be provided for as tending to overturn the structure sideways. This tendency is met by giving the legs a spread or outward batter.

Stresses will also be induced owing to the inequality of the tensions in the hoisting and lowering ropes, tending to twist the frame.

If machinery such as rock-breakers be mounted on the frame, the resulting vibration will have a destructive effect upon joints and fastenings.

Loader

The loader is a self-propelled Heavy Machinery having for main function to push and lift (Load) ground pieces. Mainly, we find LOADERS in the following industries: Construction, excavation, mining, quarry, roads and snow removal. We also find some Loaders in the handling and forestry industries.

Loader: A very versatile Heavy Equipment

One of the most popular Heavy Equipment, the loader is also very versatile and can be used for other tasks if supplied with the appropriate attachment other than his standard bucket. Thus, a loader with Forks Attachment can replace a powerful forklift truck. with grapple attachment, the Loader can be used for forestry or recycling duties.

Loader Categories

- Wheel Loader: The most popular of all, the Wheel Loader comes in several dimensions. The Wheel Loader is particularly used for tasks related to heavy construction, mining, quarry and snow removal.

- Min Loader: The Mini Loader is also a Wheel Loader but having smallest dimension. The Mini Loader is used for duties requiring less strength than his big brother, the Wheel Loader.

- Skid Steer Loader: Kind of Mini Loader, the Skid Steer comes on tracks or on wheel tires. With his special wheels system , the Skid Steer Loader provides greater on-place mobility. Therefore, this Loader is often used on terrains susceptible to produce skid.

- Crawler Loader: Kind of big Loader with tracks allowing mobility in terrains having instable shapes and consistence.

- Grapple Loader: Kind of specialized Loader used mainly for wood handling. With his grapple this Loader can also be used for refuse and residual matter sorting and recycling.

- Backhoe Loader : The Backhoe Loader is hybrid machinery combining, in less extent, Wheel Loader and Excavator functions. The Excavator is placed in rear position hence the Backhoe designation.

Winding Engine

In modern mining industry, the electric hoists have replaced the winding engines. The winding engine is known as a first motion steam winding engine. First motion, because no gearing is used, the steam pistons by way of piston rods, crosshead and connecting rods drive directly on the crankshaft which carries the rope drums.

The steam cylinders and pistons are the simple type or single expansion.

The engine is purely manually operated by the winding engine driver. The controls consist of a steam inlet valve, a reversing lever and a pedal brake on each rope drum.

Winder Operation

The basic operation of a rock winder system is to transport reef and waste from the different underground mining levels to the surface. The following components need to be defined to understand the winding process:

- Skip – a bucket in which the rock is transported.

- Winder rope – used to pull the bucket up or down.

- Winder pulley – a pulley that guides the rope up and down the shaft.

- Winder motors – drives the winding operation.

- Skip feeder – a cone like structure that loads the skip with rock.

To make the transport process easier the rocks needs to be blasted into smaller pieces. The smaller rocks are then transported to the underground silos via a mining train. Then 20 tons of rocks are loaded onto the skips via the skip feeder and this loading process takes about two to five minutes. This is then where the winding process starts.

The winder motors winds the one end of the rope and at the same time it rewinds the other end of the rope. With this winding operation we get that the one skip at the one end of the rope moves up, and at the same time the other skip at the other end of the rope moves down. The skips can reach a speed of up to 15 m/s.

When the skip arrives at the surface of the shaft, the rock is automatically thrown onto a conveyer belt that transports the ore to the gold plant.

Scraper

A scraper is commonly used in road construction, subdivision development and mining. Scrapers are heavy-duty machines that can be used for digging and hauling earth and minerals from one part of a job site to another, as well as leveling out those materials.

They typically have very large rubber tires and are often motorized – though some are tow-behind machines – and allow you to quickly and easily move product around your site to get the job done. There are four major types of scrapers, though, each offering distinct advantages for different operations.

Single-Engine Wheeled Scrapers

The single-engine wheeled scraper is perhaps the most common type of scraper. It consists of a bowl, an apron that is drops over a load of earth for transport, and an ejector that relies on hydraulics to get rid of a load once you have successfully moved it. With hydraulics, each separate function can operate separately, as well, making these exceptionally versatile machines.

Dual-Engine Wheeled Scrapers

Dual-engine wheeled scrapers are another great option if you are hauling earth for a short distance. This type of scraper has two engines, with one controlling the front wheels and the other powering the rear. This style of scraper is also highly effective for short hauls and narrow cut-and-fill areas on job sites.

Elevating Scrapers

Rather than rely on an apron like other scrapers, an elevating scraper uses an elevator that is either hydraulically or electrically driven. This elevator loads materials into a raised bowl that can then dump out a load by sliding the bowl's floor backwards, with the elevator capable of being reversed to help evenly and completely finish an offload.

Pull-Type Scrapers

Finally, pull-type scrapers are not motorized at all. Instead, these are towed behind other machines on site, but offer the advantage of being more capable of operating in wet, soft and sandy conditions. By not having its own driving capabilities, this can also make it easier to keep from getting stuck in messy terrain, which means the pull-type can be quite useful in rainy climates and springtime weather.

Rail Car

Ore Bucket Rail Cars are built of the strongest welded steel design and are built with the expectation that continuous service will be required of them.

These cars are made in two styles, both of which are built as standard units for eighteen gauge track.

However, these ore bucket cars can be promptly manufactured to any individual speciãcations desired by the customer. Style No. 2, which is equipped with bucket guard, is approximately 20% higher in price than Style No. 1. Speciãcations are given in the following table.

Style 1 Ore Bucket Car

Ore Bucket Car with Guard

Size No.	Style	Diam. of Wheels in Inches	Approx. Weight Lbs.	Size No.	Style	Diam. of Wheels in Inches	Approx. Weight Lbs.
1	1	6	175	1	2	6	210
2	1	8	190	2	2	8	225
3	1	10	215	2	2	10	250

Rocker Dump Car Ore

Rocker Dump Ore Cars may be dumped to either side and easily returned to an upright position. A locking mechanism, located at the center of the car bottom and at the center of the frame, automatically stops the body when it attains an upright position.

The body is carried on heavy cast steel rockers and rocker stands, which are covered by a plate to prevent spilled material lodging on the rockers.

The body is made of heavy plate, properly braced and are welded, making a tight container with no doors or openings through which material can leak out onto the trackway.

Rocker Dump Ore Cars are made on order to meet operating conditions. These cars are popular in capacities ranging from 20 to 40 cu. ft. and for from 18″ to 24″ gauge mine track. Write for additional information.

Rocker Dump Ore Car

Standard Car Ore

Standard Ore Car is made of the very best materials and is designed for the most severe service. Although it is regularly furnished with plain bearing wheels, roller bearings can be supplied whenever preferred. Ordinarily this car is built for standard eighteen inch gauge track although special units can be quickly manufactured to the customer's individual speciãcations wherever this is necessary.

The Standard Ore Car can be knocked down for airplane or muleback transportation to facilitate shipping to remote localities.

Standard Ore Car

Size			Size of Box in Inches			Thickness	Approx.
No.	Cubic Feet	Ave. Ore Lbs.	L	W	H	Of Steel	Ship. Weight Lbs.
00	10	1250	40	22	19	No. 16	250
01	13	1500	44	24	20	No. 14	375
02	13	1500	44	24	20	No. 12	440
03	14	1625	48	24	20	No. 10	512
04	14	1625	48	24	20	3/16"	600
05	16	1920	48	24	24	¼"	650
06	20	2400	48	30	24	¼"	700
07	23	2760	54	30	24	¼"	800

Car, Ore, (Standard Cage)

(Standard Cage) Ore Car body is of heavy plate steel and all forgings are of the best quality iron and steel. The truck frame is made of heavy channel iron with the body securely mounted. Hooks or slots are provided for securely holding the car to the cage.

The standard gauge is 18″ but other sizes can be furnished. This car can be sectionalized for airplane or muleback transportation.

911 Metallurgist

(Standard Cage) Ore Car

Size No.	Capacity	Size of Boxes, Inches			Thickness of Steel		Approx.
	Cu. Ft.	L	W	H	Sides	Bottom	Weight lbs.
C1	14	42	24	24	No. 10	3/16"	650
C2	14	42	24	24	3/16"	¼"	750

C3	16	48	24	24	No. 10	3/16"	800
C4	16	48	24	24	3/16"	1/4"	850
C5	20	48	30	24	3/16"	1/4"	900
C6	22	54	30	24	3/16"	1/4"	950
C7	28	54	30	30	1/4"	1/4"	1200
C8	32	60	30	30	1/4"	1/4"	1400

Car, Ore, (Type Z)

(Type Z) Ore Car will meet practically all conditions where cars of this style are required. The body is well braced and riveted and two angles are used to reinforce the bottom. Cars of twenty cubic foot capacity and larger are furnished with one reinforcing strap or band in the center of the body. Bumpers, used as handles when dumping, are located at the strongest part of the body. The door is well protected by a bracing strap at the bottom where the cars come together. The turntable is of cast iron with a grease lubricated machined groove, which takes most of the turning load off the king pin. The truck frame is in one piece without riveted corners to work loose or cause trouble. Brakes may be attached to any (Type Z) Ore Car.

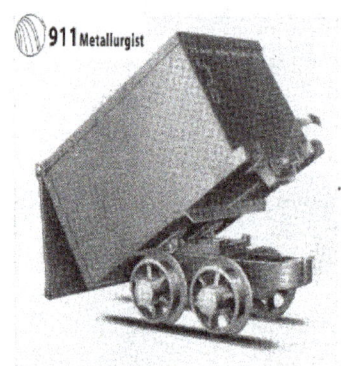

(Type Z) Ore Car

| Car No. | Capacity Cu. Ft. | Gauge Track | Dimensions | | | Thickness Material | | Ship. Wt. Lbs. |
			L	W	H	Bottom	Sides	
Z 14	14.0	18"	47"	28"	41"	3/16"	No. 10	660
Z 16	15.9	18"	49"	30"	41"	3/16"	No. 10	730
Z 20	20.0	18"	53"	34"	43"	1/4"	No. 10	1005
Z 16	15.9	24"	49"	34"	42"	3/16"	No. 10	790
Z 20	20.0	24"	53"	34"	44"	1/4"	No. 10	1075

Mine Ore Loader

The was designed and developed by practical mining men. Their wide experience in handling ore and muck in underground mines qualiãed them to design a machine which did not have all the faults and disadvantages common to mechanical loaders, scrapers, and muckers, and that this was accomplished has been proved by years of operation.

This loader, due to superiority of design, is made more powerful and faster than any other loader doing similar work. It will operate in the smallest drifts and tunnels and can be shifted with ease from one mine tunnel to another. It has a digging width of approximately 6 feet.

The Mine Ore Loader is inexpensive to operate, low in air consumption, and is constructed to withstand the roughest use; having electric treated alloy steel castings throughout, anti-friction bearings, steel gears and heavy duty air motors designed especially to stand up under the severe requirements for mine and tunnel work. Ample lubrication for all moving parts has been carefully provided. The motor reduction gear cases and transmission gear case are lubricated through easily accessible plugs. All parts are thoroughly protected against possible damage.

Mine Ore Loader Loading
(Standard) Ore Car

Gasoline Locomotive

The (Gasoline) Locomotive is an economical, powerful unit for mine car haulage where tonnage handled warrants installation of means for mechanical conveyance.

The utilization of gasoline as the motive power in haulage, both in the mining and industrial ãelds, has effected such reductions in cost as to warrant favorable consideration of this method of handling wherever practical. Some of the advantages to be obtained from mechanical conveyance by means of gasoline powered mine locomotives are low ãrst cost, ease of operation, readiness for immediate service, no expense when idle, and low operating costs. In most cases a considerable saving can be made using the gasoline powered mine locomotive in place of man or mule-powered haulage systems. The (Gasoline) Locomotive is built for maximum efãciency and will give long service under the most difãcult operating conditions and withstand considerable rough usage.

The power plant consists of a standard automobile gasoline engine of about 60 to 85 horsepower with radiator, electric generator, self-starter, clutch, gear reduction, transmission, and reversing gear. Reversing gears and shafts run in ball bearings which are completely housed. In addition to the reversing gear provided by the standard motor, which gives only one reverse speed; another is added which reverses the entire power plant. Thus, this unit is so arranged that three speeds are available in either direction. Maximum speed is 10 miles per hour.

The (Gasoline) Locomotive frame is built up of heavy cast iron members, securely bolted together and carrying the four bearings of driver axles—mounted on roller bearings. Power is transmitted to drivers through steel sprockets and roller chain.

A hand brake is provided of sufãcient power to hold any load which the locomotive will pull. This unit can be furnished for any reasonable track gauge and various hauling capacities may be obtained depending upon the percent grade where unit will be used.

(Gasoline) Locomotive

**Locomotive Weight Tons	Cap. Tons On Level	Cap. Tons 1% Grade	Cap. Tons 2% Grade	Cap. Tons 4% Grade	Cap. Tons 6% Grade
1 ½	20.0	12.0	8.5	5.4	4.0
3	40.0	24.0	17.0	11.0	8.0
4	53.0	32.0	22.8	14.5	10.6

*Hauling capacities in tons of 2,000 lbs. based on a rolling friction of 30 lbs. per ton on grades shown.

**To obtain net hauling capacities deduct weight of locomotive in tons from hauling capacities instead.

Mine Trolleys

Mine trolleys are specially equipped containers installed on the wheel pairs. They are designed for moving ore and cargo, which do not overlap mine trolleys, along underground workings and at industrial sites of mines and pits.

According to the purpose of use, they are divided into:

Cargo trolleys:

- For transportation of coal and ore along underground workings and at industrial sites, for non-ferrous metal industry, field geological explorations.

- For mine construction.

- For modification of colliery plants.

Man-riding cars:

- For transportation of people along horizontal and inclined workings. In inclined workings, man-riding cars are used both independently and in trains, consisting of the main and one or several trailer cars.

Special mine trolleys:

- For fire trains.

- For transportation of explosives, auxiliary materials, equipment, goods in containers and palletized cargo.

Underground Personnel Carrier

Where access to a mine is by a drift or inclined shaft and rail is installed, personnel transport typically utilises a vehicle permanently attached to a winding rope with provision for seating passengers. Again this vehicle is typically used as a means of attaching other rolling stock to the winding rope and, whether set-up to carry passengers or not is usually referred to as a "dolly car". A dolly car can be automatic (push button control by the passenger) or operated by an on board operator, particularly where the dolly car is utilized to haul other rolling stock into and out of the mine.

Dolly cars typically have limited capacity for passengers and additional personnel carriage(s) are attached at shift changeovers to enable the full shift to be transported in one load or lift.

Dolly car fitted for man riding as well as connecting to materials transport cars;
drivers cabin at rear; radio control via overhead aerial.

For vertical shafts, personnel transport may be in a large shaft vehicle, or "cage" with the cage being used to transport materials at other times, or may be in a special permanent personnel winder which operates more like a building lift. Most shaft conveyances are fully automated, requiring only a start button to be pressed.

There may be benefits in separating personnel and material transport systems, particularly in relation to the greater efficiency then possible of material transport. With shared systems, it is normal to suspend material transport in the event that passengers are to be conveyed in the system. Such suspensions can be extremely disruptive, particularly in large mines where the material transport system is fully utilized.

With rubber tyred systems various types of personnel transports have been developed to fit in with the materials transport (e.g, trailers or pods fitted out for personnel transport but handled by the normal materials handling prime movers). There have also been convertible vehicles developed where seating for personnel could be folded away leaving the vehicle suitable for materials. Most of these vehicles operated successfully, but interference with the materials transport system when personnel transport was required led to inefficiency, especially if some personnel transport was required for the full shift.

Purpose designed vehicles are now typically used for personnel transport such vehicles essentially being modified "landcruisers" or "troop carriers". These vehicles offer considerable flexibility as they can be used throughout the shift to carry staff, inspection personnel, tradesmen, etc who do not work in a single location and can also be quickly adapted for use as an ambulance if required. The development of the landcruiser style vehicles with improved seating and ergonomics has generally led to the demise of other types. Though several of these older vehicles are still in use they are generally being replaced by the landcruiser type when the reach the end of their economic life, if not before.

Many mines also utilize multi-purpose vehicles which can carry a limited number of people as well as materials (e.g. utility vehicles to carry tradesmen and equipment, special pipe handling vehicles, etc).

A recent development in personnel transport is the use of "non flameproof" vehicles for personnel transport in restricted areas (being intake airways). Their use in areas likely to contain gas (hazardous zones) is prevented at least by mine rules but also by attaching automatic cut-out devices operated by a radio signal from devices attached to the roadway ribs or roof. Such vehicles are much cheaper to purchase and maintain, essentially being standard off road vehicles.

It should be noted that some mines also utilise bicycles for individual transport. Provided road surfaces are in good condition they have been quite popular with mine personnel, particularly supervisors and specialist staff. However some mines have discontinued their use due to perceived safety concerns.

Haul Truck

Mining dump trucks are designed for transportation of loosened rocks on technological haul roads at open-pit mining sites under different climatic conditions. These trucks can be used in construction of large industrial structures and hydraulic facilities, in construction of highway systems as well as in technological departments of the enterprises of processing industry.

Example of Mining dump trucks

Belaz 75710

Belaz 75710, with payload capacity of 496t, is the biggest mining dump truck in the world. The ultra-heavy dump truck was launched by the Belarusian Company Belaz in October 2013 under an order from a Russian mining company. Sales of Belaz 75710 trucks are scheduled to start in 2014.

The truck is 20.6m-long, 8.16m-high and 9.87m-wide. The empty weight of the vehicle is 360t. Belaz 75710 features eight large-size Michelin tubeless pneumatic tyres and two 16-cylinder turbocharged diesel engines. The power output of each engine is 2,300HP. The vehicle uses electromechanical transmission powered by alternating current. The top speed of the truck is 64km/h.

Caterpillar 797F

Caterpillar 797F, the latest model of 797 class dump trucks manufactured and developed by Caterpillar, is the second biggest mining dump truck in the world. The truck has been in service since 2009. It can carry 400t of payload compared to its predecessor models 797B and the first generation 797, with payload capacities of 380t and 360t respectively.

The dump truck has a gross operating weight of 687.5t and measures 14.8m in length, 6.52m in height and 9.75m in width. It is equipped with six Michelin XDR or Bridgestone VRDP radial tyres and Cat C175-20 four-stroke turbocharged diesel engine. The single block 20-cylinder engine offers gross power output of up to 4,00HP. The truck uses a hydraulic torque converter transmission and runs at a top speed of 68km/h.

Terex MT 6300AC

Terex MT 6300AC, introduced by the American manufacturer Terex in 2008, is also an ultra class mining dump truck with a payload capacity of 400t. The vehicle was rebranded as Bucyrus MT6300AC, following the acquisition of the mining equipment division of Terex by Bucyrus in 2010. Terex MT 6300AC became a part of Caterpillar's Unit Rig line after Caterpillar's acquisition of Bucyrus in 2011.

The gross operating weight of the vehicle is 660t. The body of the truck is 7.92m in height and 14.63m in length. The vehicle is equipped with a four-stroke diesel engine, with 20 cylinders powering an AC electric alternator, which in turn supplies power to the electric motor fitted at each side of the rear axle. The rated power output of the engine is 3,750HP. The vehicle can move at a maximum speed of 64km/h.

Liebherr T 282C / T 284

Liebherr T 282C and Liebherr T 284 are two 400t payload capacity ultra-class haul trucks designed and manufactured by Lieb herr. The trucks share the distinction of being the second biggest mining trucks with Caterpillar 797F and Terex MT 6300AC. Liebherr's T 282C is the successor to the 360t capacity T 282B class mining truck. T 284, the latest class of trucks from Liebherr, shares many similar features with the T 282C.

The gross weight of T 282C and T 284 is the same, at 661t. The overall length of the trucks is 15.69m. The overall width and the loading height are 9.679m and 7.42m respectively.

The vehicles are equipped with a 20-cylinder diesel engine with a gross power output of up to 3,750HP. The vehicles use the Liebherr insulated-gate bipolar transistor (IGBT) AC electric drive system. The maximum travel speed of the vehicles is 64km/h.

Belaz 75601

Belaz 75601 has the capacity to haul 396t of payload. It is the latest model of 7560 class of trucks designed by Belaz for carrying loosened rocks at deep open-pit mining sites under different climatic conditions.

The gross operating weight of the vehicle is 672.4t. The Belaz 75601 measures 14.9m in length, 9.25m in width and 7.22m in height. The dump truck uses electromechanical transmission with a four-cycle diesel engine, having 20 V-type cylinders. The engine's power output is 3,750HP. The traction motors and the traction alternator are provided by Siemens and Kato respectively. The top speed of Belaz 75601 is 64km/h.

Komatsu 960E-1 / 960E-1K

Komatsu 960E-1 and Komatsu 960E-1K are the two latest rigid dump trucks by Komatsu. Each truck has a payload capacity of 360t. The 960-E1, introduced in 2008, is the first generation of the 960E series of haul trucks from Komatsu, which was followed by Komatsu 960E-1K.

The gross weight of both the trucks is 635t. The loading height and the width are 7.14m and 9.19m respectively. The overall lengths of Komatsu 960E-1 and Komatsu 960E-1K are 15.6m and 15.34m respectively. Both vehicles are powered by a four cycle diesel engine with 18 V-type cylinders. The power output of the engine is 3,500HP. Komatsu 960E-1 uses GE dual IGBT AC electric drive system, whereas the 960E-1K uses Komatsu IGBT AC electric drive system. The top speed of both trucks is 64km/h.

Terex MT 5500AC

Terex MT 5500AC, also known as the Unit Rig MT 5500AC, ranks as the one of the biggest mining dump trucks in the world. The 360t payload capacity vehicle is used for high volume surface mining.

The maximum gross vehicle weight of Terex MT 5500AC is 598t. The overall length is 14.87m, while the width and loading height are 9.05m and 7.67m respectively. The vehicle is powered by a four-stroke Diesel engine rated at 3,000HP with 16 cylinders. The vehicle uses AC electric drive system and travels at a maximum speed of 64km/h.

Belaz 75600

Belaz 75600 is currently one of the one of the largest mining dump trucks in the world. It is the first generation model of Belaz's 7560 class mining trucks designed for transporting rock-mass at deep open-pit mines under different climate conditions. The haul truck offers a payload capacity of 352t.

The maximum gross weight of Belaz 75600 is 617t. The overall length, width and loading height are 14.9m, 9.6m and 7.47m respectively. The vehicle is powered by a four-cycle turbocharged diesel engine with 18 v-type cylinders. The gross power output of the engine is 3,500HP. The traction alternator for the vehicle is provided by Kato and the traction motors are provided by Siemens. Maximum travel speed of the truck is 64km/h.

Caterpillar 795F AC

Caterpillar 795F AC has a payload capacity of 345t. The haul-truck features modular design and offers two body options including the popular MSD (mine specific design) and the gateless coal body.

The gross machine operating weight of Caterpillar 795F AC is 628t. The truck's overall length is 15.14m. The overall width and the loading height are 8.97m and 7.04m respectively. The vehicle is powered by Cat C175-16 diesel engine with a gross generating capacity of 3,400HP. The truck uses AC electric drive system solely designed and developed by Caterpillar, and runs at a top speed of 64km/h.

Hitachi EH5000AC-3

The Hitachi EH5000AC-3 has a payload capacity of 326t. It is Hitachi's latest and largest rigid frame dump truck and was introduced at MINExpo International 2012 in Las Vegas, Nevada, in 2012.

The gross machine operating weight of Hitachi EH5000AC-3 is 551t. The overall length of the truck is 15.51m. The width and loading height are 8.6m and 7.41m respectively. The truck uses the low emission Cummins QSKTTA60-CE diesel engine with 16 cylinders. The rated power output of the four-cycle engine is 2,850HP. The vehicle uses Hitachi's IGBT AC electric drive system and runs at speeds of up to 56km/h.

References

- Understanding-conveying-systems-their-importance: semcoice.com, Retrieved 17 April 2018
- Mining-machinery-trams: monacorarecoins.com, Retrieved 29 May 2018
- Scraping-the-surface-of-different-scraper-types: rackersequipment.com, Retrieved 09 July 2018
- Dumptrucks, products: belaz.by, Retrieved 17 April 2018
- Feature-the-worlds-biggest-mining-dump-trucks: mining-technology.com, Retrieved 30 March 2018

Chapter 5

Mine Safety

Safety has been a major concern in the mining industry. Mining ventilation, gas ignition, cave-ins and rock falls, heat strokes, etc. are some of the hazards in mines. The varied ways of ensuring safety in mines, such as through the implementation of personal protective equipment, self-contained self-rescue device, good mine ventilation systems, safety lamp, etc. have been elaborately discussed in this chapter.

Mine safety refers to the management of operations and events within the mining industry, for protecting miners by minimizing hazards, risks and accidents. Most of the safety issues related to mining are addressed in the relevant laws, compliance and best practices that are to be considered for the best possible protection of the mining workers. Employers are to abide by the laws and practices to ensure the maximum observances of safety.

The following topics are typical when discussing mine safety:

- General safety - general aspects of safety which are common to all types of mines (electrical and machine safety).

- Occupational safety and health - Issues particularly associated to the mining. These include: blasting explosives, ergonomics, diesel and dust control and hearing loss etc.

- Process and production safety - Safety within the processes associated with mining.

- Workplace safety - Safety issues directly related to the workplace (Ex. ventilation).

- Fire and explosion safety - In particular, the risks associated to fires and explosions in the mining industry.

- Structural safety - Safety in mine construction and geologic characterization.

- Environmental safety - Issues of environmental safety (direct or indirect impact of the mining industry).

Mining Safety

The mining industry is not without its risks, so workers need to be aware of mining safety tips that might save their life. Although there are industries with higher injury rates, the injuries incurred in mining are far more likely to be severe than those incurred by workers employed in private industry as a whole. Bituminous coal underground mining employs more than half of all miners in the US and experiences a higher share of occupational injuries, illnesses, and fatalities, compared to other industries. According to the most recent report from the Bureau of Labor Statistics, the rate of fatal injuries in the mining industry in 2015 was 9.8 per 100,000 full-time workers the vast majority of which come from coal mining, and is almost three times the average rate in the private sector.

This figure is actually much lower than it was even just a few years ago. In 2006, the fatality rate in mining was 24.8 per 100,000, making it one of the most dangerous industries in heavy industry. The 26 deaths in 2015 was the lowest figure ever recorded in the US and is proof of the vastly improved safety standards and training in the last decade. However, this fact is little solace to the families of the 26 workers who did lose their lives in 2015. Mining is still a dangerous job and the industry needs to keep striving to improve safety for its workers.

Common Health and Safety Hazards in the Mining Industry and some mining safety tips to help improve safety

The mining environment poses significant health risks to the workers employed in the industry. The seven most common hazards in the mining industry are:

- Chemical hazards

- Dust hazards

- Heat stress

- Musculoskeletal disorders

- Noise

- Whole body vibration

Chemical Hazards

Mine workers are exposed to a number of chemical hazards in the course of their job. There is often a chemical separation process in mining, where the metals and minerals being mined are separated from an ore/substance. Polymeric chemicals are used in the flocculation process in coal mining to treat waste water by making particles clump together and float to the top, making them easier to be removed. Chemicals are huge potential risks to workers health, with burns, poisoning and respiratory problems all risks related to exposure to chemicals.

When dealing with chemicals, there needs to be a standard operating procedure (SOP) implemented and included in all training. Different chemicals are used depending on the mining project, so there needs to be site-specific training that takes into account the chemicals used. Mining Safety Tips include providing Personal Protective Equipment (PPE)s to all workers who may come in contact with chemicals, with safe handling procedures established throughout the company.

Dust Hazards

Dust is one of the most significant hazards in mining. Coal mining can result in large amounts of airborne dust particles. The most severe risk involved with dust in coal mining is coal workers' pneumoconiosis (also known as black lung disease), which is a disabling and often fatal lung disease. In metal, nonmetal, stone, sand, gravel and some aspects of coal mining, silica dust can be released into the air and inhaled by mine workers. This can lead to silicosis, a lung disease which has led to the death of more than 14,000 workers in the past 50 years.

Dust hazards need to be closely controlled, with a hierarchy of controls implemented to protect worker health. Where possible, the dust-related hazard should be eliminated completely. If this is not possible, the dust should be substituted with a safer alternative, or the hazard isolated so as to separate workers from risk. These are the best mining safety tips to keep workers safe. If the dust hazard is unavoidable, tools and equipment can be modified with safety in mind, PPE needs to be mandatory and health and safety training and procedures need to be implemented to reduce the risk to workers.

Heat Stress

Mining workers are especially susceptible to heat-related injuries and illness. Mines are enclosed spaces that can become incredibly hot, given the lack of natural air and the physical exertion of the work involved. Risks to workers exposed to high temperatures include dehydration, heat stroke and heat exhaustion. Workers may find it hard to concentrate which can lead to injury, and untreated heat stress can result in death. To keep everyone safe, there may need to be extra rest breaks given to workers operating in high-temperature environments. Keeping workers hydrated needs to be a priority, as well as providing PPE and work clothes appropriate for hot conditions.

Musculoskeletal Disorders

Musculoskeletal disorders (MSDs) affect the muscles, nerves, blood vessels, ligaments and tendons. Workers are at risk from MSD's through actions such as lifting heavy items, bending, reaching overhead, pushing and pulling heavy loads, working in awkward positions. MSD's account for 33% of all worker injuries and illnesses.

MSD's are a significant risk in mining, given the strenuous physical activity that the work involves. Mining safety tips include training workers in the correct ergonomic procedures so that they work as safely as possible. Workers should be involved in the process to encourage early reporting of MSD symptoms, which helps catch injuries before they become serious.

Noise

Mines are noisy places with equipment and drilling reverberating to create a very loud work environment. Overexposure to high volumes of noise, over an extended period of time, can cause workers injuries such as tinnitus, concentration problems and even deafness. Where possible, excessive noise should be eliminated. However, heavy equipment is usually necessary in mines, so noise is unavoidable to some extent. Regular maintenance of equipment can help reduce noise levels and all workers should be given proper PPE and health and safety training to protect themselves from noise.

Whole Body Vibration

Any worker who uses heavy equipment is at risk from whole body vibration (WBV). Repeated use of machinery, or operating machinery in awkward positions can lead to WBV. Symptoms include musculoskeletal disorders, reproductive damage in females, vision impairment, digestive problems and cardiovascular changes.

WBV is a serious hazard and safety measures should be implemented to keep workers safe. There need to be regular breaks for those using heavy machinery, as well as training so that workers are aware of the risks, and can flag them early, before any injuries are sustained.

Despite significant progress in recent years, the mining industry is still one of the most dangerous workplace environments. When safety protocols are ignored for the sake of getting the job done faster, the effects on human health can be dangerous and often lethal. With so many different hazards, employers must take no chances and implement as many safety regulations as possible to protect their workforce.

Mine Ventilation Systems

Ventilation is the control of air movement, its amount, and direction. Although it contributes nothing directly to the production phase of an operation, the lack of proper ventilation often will cause lower worker efficiency and decreased productivity, increased accident rates, and absenteeism.

Air is necessary not only for breathing but also to disperse chemical and physical contaminants (gases, dusts, heat, and humidity). In the U.S., as well as in the rest of the world, mine ventilation practice is heavily regulated, especially in coal and gassy (noncoal) mines, and other statutes relate to air quantities required to dilute diesel emissions, blasting fumes, radiation, dusts, battery emissions, and many other contaminants.

To ensure adequate ventilation of a mine, provision is made for suitable paths (airways or aircourses) for the air to flow down the mine to the working places and suitable routes out of the mine when it has become unsuitable for further use. The primary ventilation system thus consists of an intake or intakes (or downcasts) through which the fresh air passes, the mine workings, and an exhaust or exhausts (or upcasts) where the air passes after having ventilated the working places of the mine. Mine fans can be installed on the intake airshaft, return airshafts, or both, either on the surface or underground (figure).

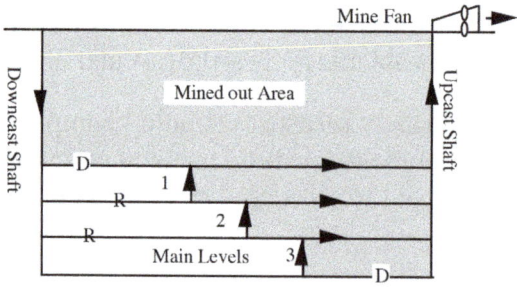

Figure: Basic ventilation system underground where D is a ventilation door or airlock, R is a mine regulator and 1, 2, 3 are working places with a surface exhaust fan.

To maintain adequate ventilation through the life of a mine, careful advance planning is essential. Advance ventilation planning involves the consideration of two principal factors: (1) the total volume flow rate of air required by the mine, and its satisfactory and economic distribution, and (2) the pressure required by the mine fan(s). A well designed ventilation system should be effective, flexible, and economical.

Mine System and Control Devices

A well designed and properly implemented ventilation system will provide beneficial physiologicaland psychological side effects that enhance employee safety, comfort, health, and morale. In planning a ventilation system, the quantity of air it will be necessary to circulate to meet all health and safety standards must be decided at the outset. Once the quantity required has been fixed, the correct size of shafts, number of airways, and fans can be determined. As fresh air enters the system through the intake airshaft(s) or other connections to the surface, it flows along intake airways to the working areas where the majority of pollutants are added to the air. These include dust and a combination of many other potential hazards, such as toxic or flammable gases, heat, humidity, and radiation. The contaminated air passes back through the system along return airways. In most cases, the concentration of contaminants is not allowed to exceed mandatory threshold limits imposed by law. The return (or contaminated, exhausted) air eventually passes back to the surface via return airshaft(s), or through inclined or level drifts.

Air always flows along the path of least resistance, but this may not be where it is required for use. To direct the air where it is needed, ventilation devices are necessary; the primary means of producing and controlling the airflow for the entire system are mine fans (either in the form of single fan installation or multiple fans). In addition, many other control devices also are necessary for effective underground air distribution:

1. Stoppings - Temporary or permanent

Stoppings are simply air walls made of masonry, concrete blocks, pre-fabricated steel, gob walls, fire-proofed timber blocks, or any other material used to channel airflow for effective air distribution. Depending on the size of mining entries, stopping sizes range from as small as 4-ft by 20-ft in low coal to as large as 30-ft by 40-ft in limestone mines.

2. Overcast/Undercast

Overcasts are air bridges where intake and return airways are required to cross each other without mixing. They could be constructed of masonry, concrete blocks, or pre-fabricated steel.

3. Regulator

Regulators are commonly used to reduce the airflow to a desired value in a given airway or section of the mine. Depending on its permanency and the pressure differential to be experienced across the regulator, materials used in the construction of regulators range from a simple brattice sheet blocking the airway to a sliding shutter in a stopping.

4. Man-doors

These are generally steel access doors mounted in stoppings between intake and return airways.

5. Air locks

When access doors between intake and return airways are necessary and their pressure differential is high, man-doors generally are built as a set of two or more to form an air-lock. This prevents short-circuiting when one door is opened for passage of vehicles or personnel. The distance between doors should be capable of accommodating the longest train of vehicles required to pass through the air-lock.

6. Line brattice/Vent tubing

As a short term measure, fire-resistant line brattices may be tacked to roof, sides, and floor in underground coal mines to provide temporary stoppings where pressure differentials are low in and around working areas. For metal and non-metal mines, vent tubing is generally used in and around working areas to channel fresh air to operating faces. Vent tubing is also commonly used in combination with auxiliary fans.

7. Booster/Auxiliary fans

When the airflow in a section of the mine must be adjusted to a magnitude beyond that obtainable from the open system, a booster fan may be used to enhance the airflow through a part of the mine. When they are used, they should be designed into the system in order to help control the leakage, without causing undesirable recirculation in either normal or emergency situations. In the U.S., booster fans are prohibited in underground coal mines.

8. Machine-mounted watersprays and scrubbers

These are devices used to enhance the flow of fresh air in face areas. Scrubbers are "vacuum cleaners" used for dust suppression, while watersprays, when strategically located on machines, have been used successfully to act as a "booster fan" to re-direct airflow in certain directions in face areas.

Major Ventilation Systems

The objective of any ventilation system is twofold. First, the primary ventilation must course air through the main airways to the immediate working area outby the working faces, thus making fresh air available for face ventilation, and then return the contaminated air through return (exhaust) airways to the surface. Second, the face ventilation system must be designed to effectively utilize the available air in the immediate working area to sweep the working face, to capture and remove dust, and to dilute and carry away gas, if any, emitted during mining activities. Without a properly designed ventilation system, an efficient production cycle would not be possible. The system should provide the required air volumes and quality at reasonable pressure losses, perform with minimum interference and cost to production, and do so in the most cost-effective way possible. Furthermore, the primary ventilation system may be well designed, but if the available air brought to the working area is not properly utilized for ventilating the faces where most workers are located, the total system has failed.

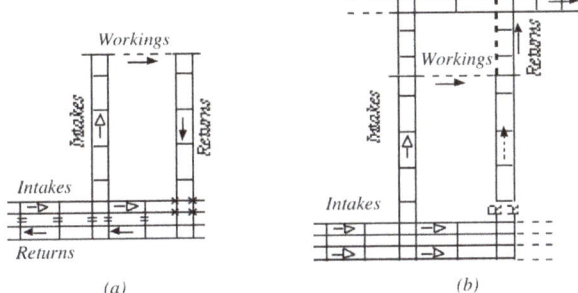

Figure: Basic ventilation systems (a) U-tube and (b) through-flow

Depending on the type of mine and disposition of local geology, ventilation layouts can be divided into two broad classifications; either a U-tube system or a through-flow arrangement. Figure shows a basic U-tube configuration where air flows towards and through the working area, then returns along adjacent airways, often separated from intakes by long pillars and/or stoppings. Access doors in the stoppings facilitate traffic between intake and return airways. The variation of this arrangement would be room-and-pillar and longwall type mining methods. The other arrangement is shown in Figure, where intakes and returns are usually separated geographically from adjacent airways, which are either all intakes or all returns. Although fewer stoppings and airways are needed because of the geographical separation, which often results in less air leakage, air current regulations and boosters may be required for airflow control in work areas. Parallel flows between intake and return airshafts across the multilevel metal mines and the bleeder system in a longwall panel would be typical examples of this type layout.

Actual layouts underground could be variations of any one system or a combination of the two arrangements.

For Stratified Deposits

The vast majority of underground mines extracting tabular forms of orebodies (coal, potash, salt, limestone, etc.) normally use one of two methods, longwall or room-and-pillar mining. While actual layouts can vary significantly from mine to mine and region to region according to local geological conditions, the basic design for these two methods remains the same. The following sections describe the airflow distribution system usually employed.

a) Longwall systems

Two factors that have significantly influenced the design of the longwall ventilation systems are the control of methane or other gases that accumulate in the gob area and the increasing high rate of rock breakage on heavily mechanized longwalls that has exacerbated the production of dust, gas, heat, and humidity.

Figure depicts some of the commonly used ventilation layouts used on longwall sections. In the U.S., a minimum of two entries is required, while single entry longwalls are primarily employed in European coal mines.

System layouts become more complex when mining under inclined, thick, and gassy coal seams with frequent faults. Narrower and shorter panels are necessary to cope with these difficult conditions. There also have been other type of layouts to accommodate specific geological conditions.

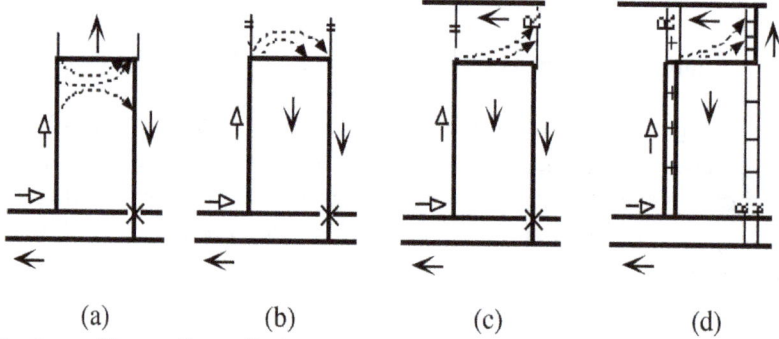

(a)   (b)   (c)   (d)

Figure: Classifications of longwall ventilation systems (a) single-entry advancing; (b) single-entry retreating; (c) single-entry retreating with back bleeder; (d) doubleentry retreating with back bleeder

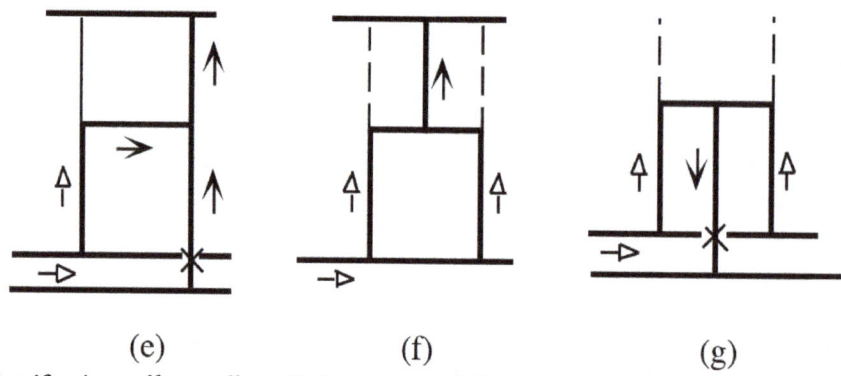

(e)   (f)   (g)

Figure: Classifications of longwall ventilation systems (e) Y-system; (f) double-Z system; (g) W-system.

b) Room and Pillar Systems

Figure shows the two methods of ventilating a room and pillar development panel in a coal mine where multiple entries are driven. The directional, or W-system, in which intake air courses are airways in the central portion of the panel, with return airways on both sides, often referred to as the fish-tail method. ThIS method is the unidirectional system in which intake and return are located on both sides of neutral airway.

In both cases, the conveyor belt and/or track are located in the middle, with a brattice curtain at the end to regulate the airflow. In U.S. coal mines, unless a special petition is approved by MSHA in advance, air in these entries is not supposed to be used to ventilate working areas under normal circumstance, so they are directed directly to the return airway through a regulator. Advantages of this system include: the airflow splits at the end of the panel, with each airstream ventilating the operational rooms sequentially over one half the panel only, resulting in less leakage due to less pressure differential across the stopping; and any gas emission will be flowing automatically to return airways. An obvious disadvantage is that the number of stoppings required is double that of the uni-directional system. The air leakage also is twice as much due to the extra stoppings.

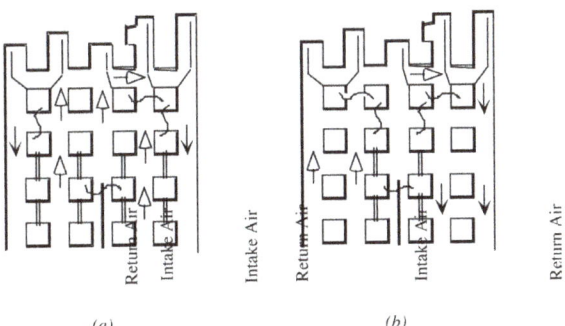

Figure: Room and pillar development with line brattices to regulate airflow in conveyor belt entry: (a) bi-directional system; (b) uni-directional system.

A uni-directional system should offer a higher volumetric efficiency at the face because of the reduced number of stoppings. A disadvantage is that the higher volume also has a higher ventilation pressure, which in turn offers higher leakage.

c) Mine with Large-size Entries

Typically, mines with large-size entries (e.g., limestone, salt, and oil shale) require large volumes of ventilating air (between 350,000 to 500,000 cfm, depending on specific conditions) to adequately ventilate underground workings. In trying to meet this requirement, two major problems are usually encountered: (1) excessive air leakage through stoppings and (2) local air recirculation, both of which are caused by improperly constructed (and maintained) stoppings, or the lack of stoppings in many cases, and both can adversely affect the underground working environment. Oftentimes, mine management are reluctant to construct an adequate number of stoppings, either because of technical problems or the associated expenses.

Air is used to dilute diesel exhaust and to maintain a minimum air velocity in large-dimension airways in order to avoid air stratification, and every reasonable measure has to be taken to ensure that fresh air is effectively delivered to working places where air is needed. The cost of not maintaining adequate ventilation results in a poor working environment that not only is in violation of federal and state regulations, but can adversely affect worker performance and morale. To deliver fresh air to working places over large distances, effective air distribution system is essential. They can either be: (1) conventional large-scale stoppings using pipes with metal sheeting, brattice and wire, a muckpile, etc. or (2) adopting a modular type pillar layout.

Constructing air-tight stoppings in large openings is not only time-consuming and expensive, it often is difficult, if not impossible, to be 100% effective. The precise cost of constructing conventional metal-frame stoppings in a 35-ft wide by 20- ft high entry is difficult to obtain because of the many variables involved, e.g., ranging from $20,000 to $24,000 per stopping in a limestone operation in Iowa.

Oftentimes, brattice curtains are the only practical materials for use underground; the cost of which ranges from $1,500 to $3,000 per stopping (1997 dollars). This includes the cost of labor and materials. However, stoppings such as these are subjected to much higher leakage between the strips of brattice and around the peripherals. Leakage varies depending on many factors, such as workmanship, maintenance, mining practices (stoppings too close to working areas will suffer frequent blasting damage), and, to a lesser extent, geological conditions (roof sagging and bottom heaving can damage stoppings). A salt mine in Ohio with 20-ft x 40-ft entry sizes showed a leakage rate of 5,100 to 5,500 cfm per stopping.

Some mines have hung a continuous brattice line along the pillars, which stops most the leakage around the peripherals (figure).

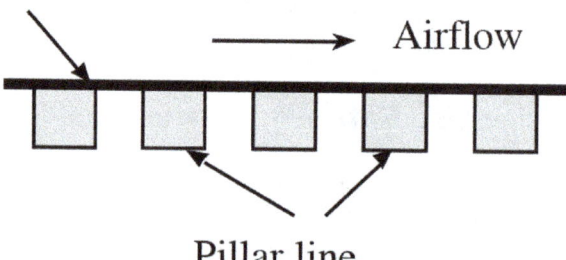

Figure: A continuous brattice line will reduce leakage around peripherals.

Although the exact impact on power cost due to leakage is difficult to quantify, it is known to be significant. Since any leakage through a stopping has to be compensated for by "pumping" more air underground to meet safety requirements, it will dramatically increase energy requirements at the fan because fan power and air quantity have a cubic relationship. For example, a 26% increase in air flow would double the air power cost. Costs of other types of stoppings also can be estimated roughly in today's dollars using published information.

A modular configuration provides an alternative. In this layout, long barrier pillars are intentionally left at four sides of a pre-planned mining block so air can be effectively coursed over longer distances. The following diagram shows a hypothetical mine working with a modular configuration.

There will be small losses in percentage extraction. However, this loss can be reduced to a minimum by leaving only the last round of mining in a cross-cut (figure). The reduction in air leakage, plus the savings created by replacing brattice and providing effective ventilation, will offset the cost associated with a lower extraction ratio. The following diagrams show the extraction ratio calculations. Figures show 40-ft by 40-ft pillars mined on 70-ft centers. Calculations show that the difference in production is approximately 6%. But a more realistic figure would be around half that, or 3%, because only a fraction of the partial pillars are left on four sides within this block. Figures show a similar 70-ft center pillar pattern, however, 35-ft by 35-ft pillars are left. The difference caused by leaving a partial pillar is around 3.5%.

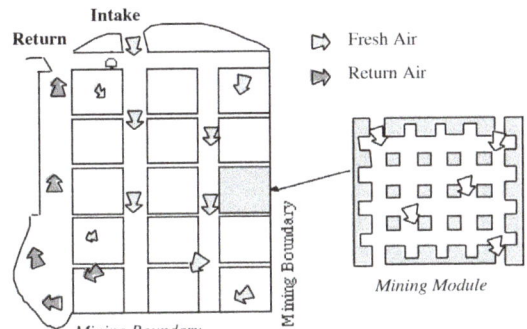

Figure: A hypothetical underground limestone mine where a module system is used in lieu of stoppings to channel air to working areas.

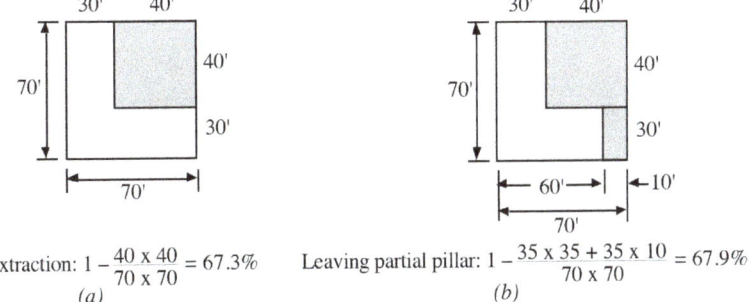

Full extraction: $1 - \frac{40 \times 40}{70 \times 70} = 67.3\%$ Leaving partial pillar: $1 - \frac{35 \times 35 + 35 \times 10}{70 \times 70} = 67.9\%$

(a)   (b)

Figure: Extraction ratio calculations.

$$\text{Full extraction} : 1 - \frac{35 \times 35}{70 \times 70} = 75.0\%$$

$$\text{Leaving partial pillar} : 1 - \frac{35 \times 35 + 35 \times 10}{70 \times 70} = 67.9\%$$

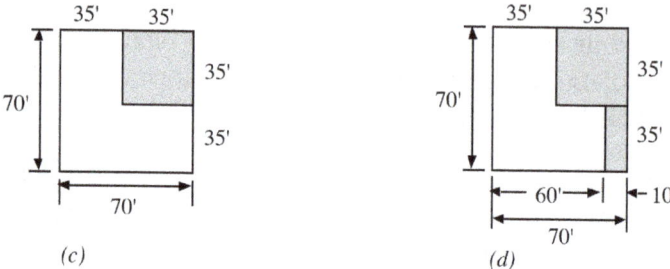

Figure: Extraction ratio calculations.

For demonstration purposes, in a 6-pillar by 6-pillar block with 20-ft room hight by 35-ft entry widths, partial pillars will be left in 16 places in figure below. Assume the stones has a density of 165 lb/ft³, total unmined rock in this system amounts to:

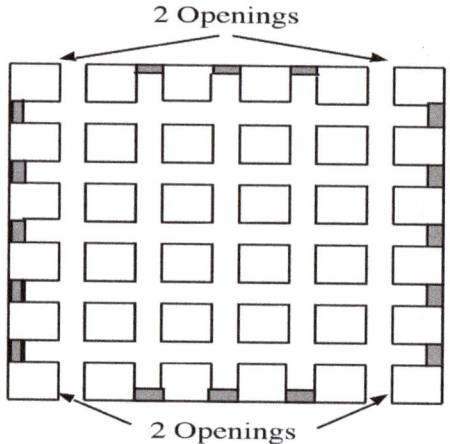

Figure: Partial pillars will be left in 16 places in a 6 x 6 block.

35' x 20' x 10' (height) = 7,000 ft³ per working place

Total tons lost per working place = (7,000 ft3 x 165 lb/ft³) ÷ 2,000 lb/ton = 578 tons/place

Total tons lost due to partial pillars = 578 tons/place x 16 places = 9,248 tons

d) Recirculation of Air Underground

In addition to improperly constructed and poorly maintained stoppings, the other major problems underground is air recirculation which is caused by improper mine layout and lack of adequate stoppings. One of the commonly encountered planning errors in underground limestone operations is that main intake and return airways are located adjacent to each other, causing the exhaust air to be recirculated back into intake airway(s). This situation is exacerbated when a box cut, often used in limestone and coal mines where ore crops out, is used to enter the mine. The air can not be discharged away from the intake area.

At the early stage of the mine development, main intake and return airways are usually located near each other, and it will be at least several years before an airshaft can be drilled some distance from the portal and an entire ventilation circuit can be completed. It is recommended that intake and return in the portal areas be physically separated in the start up of the mine to avoid air recirculation (figure).

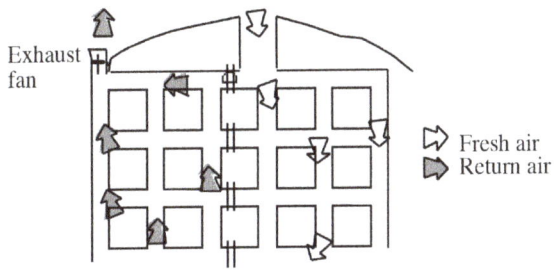

Figure: Fan duct is used to deflect exhaust air and to recover lost velocity pressure.

If this is not possible, and both intake and return have to be located next to each other at the bottom of the boxcut, Figure below shows an alternative method of separating intake air and return air by the use of a vertical fan duct at the discharge end of the exhaust fan.

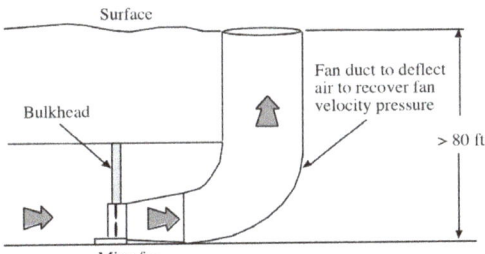

Figure: Fan duct is used to deflect exhaust air and to recover lost velocity pressure.

Figure shows a suggested layout where exhaust air is deflected upward using a vertical duct (evasè) at least 80 feet in length. The evasè also serves another useful purpose: it can recover fan velocity pressure (ranging from 0.2 in. to 1 in. W.G. depending on fan discharge velocity) which otherwise would be lost. To avoid additional shock losses, the connecting bend should be round and smooth.

Local recirculation occurs primarily because of (1) inadequate number of or leaky stoppings or (2) non-bulkheaded auxiliary fans. As a result, air will move in and out of return areas on the other side of stoppings, or circles around the auxiliary fan without going to working faces. Constructing adequate number of stoppings and properly locating auxiliary fans are primary means for reducing local recirculation.

Some of the other characteristics for mines with large openings are their low air velocity in airways which can result in air stratification, or fogging, in airways caused by influx of ground water and seasonal moisture content fluctuation. Excessive and fluctuating moisture content in air can also contribute to the deterioration of roof layers, causing safety problems. Also, for limestone operations, the combination of grade variation and fluctuating market demand requires certain flexibility in mine development, which also will impact ventilation planning.

Air Tempering for Roof Stability

In a mine with large openings, in the summer and sometimes in late spring and early fall, as warm, humid air enters the mine entries, air velocity is greatly reduced due to these large openings. The air starts to interact with the surrounding rock and quickly reaches mine temperature. As air temperature is reduced, its relative humidity will rise until the air is unable to carry the high moisture levels. Moisture at that point begins to condense on, or be absorbed by, surrounding rocks. The intake air continues to cool and dry along the length of the entries until an equilibrium is reached between mine temperature and specific humidity levels.

Laboratory studies have shown that very short term (daily) fluctuations in specific humidity have little or no effect on shale moisture gains or losses. Shales require a 7- to 10-day exposure to changes in specific humidity, whether higher or lower, before equilibrium between moisture content in the rocks and atmospheric conditions is reached. As the intake air travels along the entries, contact is made with rock and other materials in disequilibrium. Gradual moisture exchange takes place from that point at increasing residence time until equilibrium is again achieved. Figures show the seasonal effects (humidity and temperature) for an earlier study in a coal mine where the roof is predominately shale.

Moisture exchanges between incoming air and surrounding rocks are site specific. Although the exact moisture absorbing characteristics of the rock are not known, studies on shale in other mines can be used as a guideline to reasonably estimate their interactions. Previous USBM research indicates that the maximum efficient air residence time for tempering summer air is 30 minutes; additional residence time will provide total tempering. In the above mentioned study, 60 grains per pound of dry air is assumed as a good approximate summer moisture equilibrium value and its results showed a 70% reduction in excess moisture in 15 minutes residence time. It was estimated that, if the residence reached 30 minutes, moisture reduction could be expected to reach 90 to 95% tempering. Based on this, a minimum of 15 minutes residence time can be used as a good design parameter that will provide sufficient tempering effect. Having adequate air velocity is the most effective means for reducing moisture condensation.

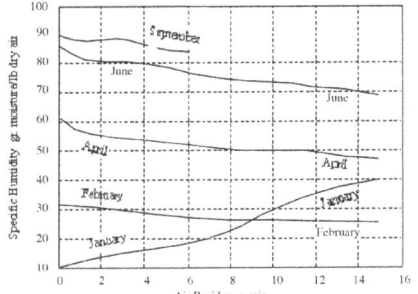

Figure: Changes in mine air specific humidity as a function of air residence time

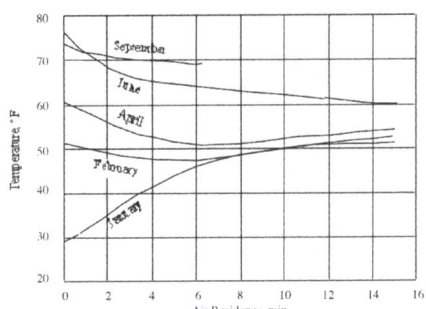

Figure: Changes in mine air temperature as a function of air residence time

Orebody Deposits

Metalliferous orebodies often occur in deposits of irregular geometry, varying from tortuous veins to massive irregularly shaped deposits of finely disseminated metal and highly variable concentrations. This causes mining layout necessarily less ordered than those for stratified deposits. In addition, the grade variation in metal mines and ever-changing metal prices necessitate that more stopes or working places be developed than would appear to be necessary, while with perhaps only a fraction of them operating on any one shift. Thus, the ventilation shift must be sufficiently flexible to allow airflow to be directed wherever it is needed on a almost day-by-day basis.

Ventilation networks for metal mines tend to be more complex than for stratified deposits and usually are also three dimensional. Figure illustrates the ventilation plan of many metal mines, although the actual geometry will vary widely.

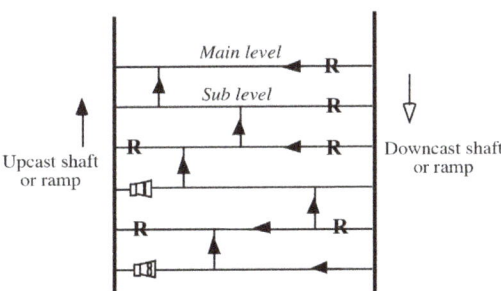

Figure: Section showing the ventilation system for a metal mine.

Airflow distribution systems for individual stopes also are subject to great variability, depending on the geometry and grade variations of the orebody. In most of the cases, where controlled vertical movement of the air is required, stope airflow systems employ ascensional ventilation. Although auxiliary fans and ducts may be necessary at individual drawpoints, every effort should be made to utilize the mine ventilation system to maintain continuous airflow through the main infrastructure of the stope. Series ventilation between stopes should be maintained so that blasting fumes can be cleared quickly and efficiently.

With respect to fan locations and airflow direction, there are primarily three ventilation systems: exhaust (pull) system, where the mine fan is located on top of the return airshaft; blowing (push) system, with the mine fan installed at the intake airshaft; and combined system (push-pull), with fans on both the intake and return airshafts. This refers to main ventilation systems only; local arrangements, for example a face ventilation system for a working area, can be different from the main system.

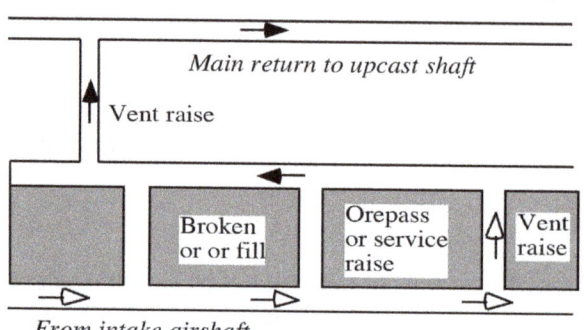

Figure: An example of a simple ventilation system for shrinkage or cut-and-fill stopes.

Depending on the particular system, the mine pressure could be either negative (exhausting system, since the fan creates a suction in the system, putting mine pressure below the atmospheric datum) or positive (blowing system). This is because the mine pressure is measured against atmospheric pressure, as shown in figure:

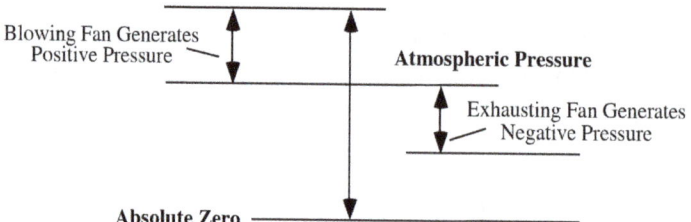

Figure: Schematic showing a positive mine pressure for blowing system and a negative mine pressure for exhausting system.

The best way to describe the relation between pressure loss and air traveling distance is to use a pressure gradient. The coal mine in figure shows that pressure loss is zero at the surface, slowly building up to approximately 0.2 in. W.G. as it proceeds 2,000-ft underground. Pressure drop accumulates to 0.7 in. W.G. after 9,000-ft just before the longwall section. Total mine pressure is – 6 in. W.G. at the end of the circuit. The negative sign simply means it is an exhaust ventilation in which pressure is lower than the atmospheric pressure.

Airflow Direction

Airflow direction is affected by the location of the main fan which, in turn, will significantly impact the other aspects of operation or transportation. An antitropal system is one in which the airflow and transported rock move in opposite directions, implying that the mineral transportation is carried out in intake airways. This tends to put restrictions on the air velocities in intake airways so that dust and other gases will not be too excessive. Conversely, a homotropal system is one where the airflow and the mined rock move in the same direction, or the haulage is carried out in return airways. This system will ensure that dust, heat, and other pollutants from broken rock will be vented directly to the outside. In addition, this system also has the advantage in case of fire occurring in the haulageway.

Another factor in airflow direction is the inclination of the airway. An ascensional ventilation is when the airflow moves upwards through inclined workings. This takes advantage of the natural ventilation effects caused by the added heat to the air. Descensional ventilation may be employed on a more compact system, with both air and conveyed materials moving downhill.

It is the pressure difference that causes air to flow underground, regardless of how this pressure difference is generated. There are different merits and drawbacks in each system. A particular system must be selected to accommodate the specific mining situation, and not the ventilation designer's pet theories. The following is a list of pros and cons for both arrangements:

1) Exhausting System

Advantages:

When main fan stops, underground pressure builds up to atmospheric. The increase in pressure slows gas emissions from the gob and prolongs the time required for the gas to reach active workings.

The haulage roads, where most travel is done, are kept free from dust, gas, and smoke. This permits the men to perform their work in fresh air.

In the event of a fire or explosion, exhausting ventilation enables the rescue work to proceed more rapidly, because the fresh air is on the haulage road, which provides an easy route for carrying material and equipment to make mine repairs.

Both intake airways and track entries serve as escapeways, if stopping lines are well maintained.

Greater power savings are possible if mine openings are small. This is due to the potentially greater recovery of velocity pressure through the use of discharge evasè (gradual expansion ducts) on exhaust fans.

Disadvantages:

It reduces temperatures in the belt slope, slope bottom, and main haulage line. During winter, the belt sprinkler system, damp coal on the belt rollers, and water lines along the haulageway can freeze. The temperature also is uncomfortable for the people working in these areas.

It is more difficult to detect a fire in belt and track entries since the air is carried directly to the return airways.

Dust produced at the portals and along the haulage road contaminates the intake air stream. Similarly, fire in the belt and track entries can be carried to the working areas.

Contaminated air goes through the fan, corrosive particles settle on the fan blades and corrode them, reduces effective air passage area, and can throw the fan out of balance.

2) Blowing System

Advantages:

It applies a continuously decreasing overpressure from the air intake portal to the discharge opening. This characteristic produces airflow from intake airways to the return and prevents contaminated flow into working areas from idle areas and return airways. In fact, the blowing system may be the only practical method of ventilation in shallow mines having fractured ground, as well as areas of contiguous mining where there may be ground cracks into abandoned mines.

The haulage roads and hoisting shaft stay free from ice, making it more comfortable for the men in winter.

A fire in any part of the mine is soon evident, due to leakage, to anybody working in the air current coming away from the face area.

Only outside air, non-corrosive and with normal moisture content, goes through the fan.

Fan unit is cheaper because of a shorter fan duct (diffuser).

Disadvantages:

Products of combustion from a mine fire or explosion are carried into the neutral escapeway. Thus, fire-fighting and rescue work are more difficult because access is often blocked by smoke. Ventilation reversal, in these cases, may endanger the men.

Dust, smoke, and other impurities are carried away from the face area and along the haulage road. Methane tends to accumulate in pockets in the roof, sometimes causing slight explosions.

Since neutral air flows away from working sections to the slope bottom, any accumulated air contaminants converge on workmen in the slope bottom area.

Shock losses are greater. It requires a distance of 30 times the duct diameter away from the pressure jet for the air velocity to lose 90% of its original velocity. For an exhausting system, only one duct diameter distance is required to lose 90% of its velocity. As result, pressure loss caused by shock, which is in addition to frictional loss, is considerably more in a blowing system.

Dirt and dust from outside will settle on the fan blades.

3) Push-pull (combination) System

In the push-pull system, it is easier to get air to difficult places. The disadvantage of this system is that it is harder to balance the ventilation system, resulting in neutral spots in the mine. According to recent survey, the majority of the underground coal mines in the U.S. use exhausting ventilation as their main ventilation system.

Ventilation Planning and Designing

The ultimate goal for ventilation planning is to design a system that will be capable of adequately ventilating all working faces, airways, and areas underground at minimum costs. A good mine ventilation system always begins with the initial development of the mining plan, which should always have alternatives. A well-thought out ventilation system can minimize long-term problems, builds in flexibility for expansion without exorbitant cost, reduces up-front capital expenditures, and phases in capital outlay over the life of the project.

Air volume requirements can be substantial in some operations. The presence of diesel, gaseous products, strata gases, heat and humidity, and large openings all require a significant increase in the minimum air velocity required, and hence a higher air volume requirement. Since energy requirement is proportional to the cube of air volume circulated, optimization between air volume and resistance must be considered. Other factors also must be factored in, such as environmental requirements and available resources. Figure shows a basic model for planning a new underground mine.

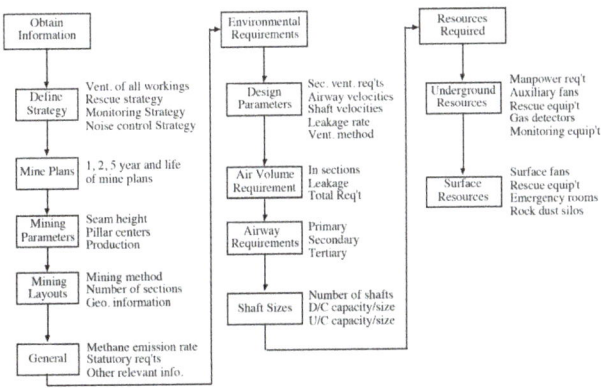

Figure: Factors considered in planning a new mine

Although many factors enter into an ultimate ventilation planning scheme, minimizing friction and shock losses are the two most important among all the items considered.

1. To minimize friction losses

From the equation,

$$R = \frac{KOL}{5.2\,A^3 N^2}$$

anything that can lower K, O, and L and increase A and N will reduce R, and ultimately lower costs in terms of having lower overall ventilation pressure, realizing that there will be practical limitations.

2. To minimize shock losses

Since up to 30% of total losses underground comes from shock losses, it is essential that shock losses be minimized to lower costs. From the equation,

$$H_x = XH_v$$

The obvious thing to do is to lower shock loss factor, X, for given air velocity. This can be achieved by rounding off corners, avoiding sudden air velocity and airflow direction changes, etc.

Ventilation Planning and Design Parameters at Homestake

The goal of a ventilation system is to provide a work environment that contains minimal safety and health risks and is conducive to hard work. Since ventilation is a cost function, this goal should be met as inexpensively as possible.

Homestake is the largest underground gold mine in the U.S., currently operating at the 8,000- foot level. The mine is ventilated by 1,069,000 cfm (504 m³/sec) of air measured at mid-exhaustcircuit density. Its air conditioning system includes a 2,300 ton (8.1 MWR) controlled recirculation plant, a 580 ton (2.0 MWR) chilled water plant, a 290 ton (1.0 MWR) drite exploration plant, 28 spot-coolers totaling 960 tons (3.4 MWR), and 35 spray coolers totaling 420 tons (1.5 MWR). The mine employs 117 diesel units with a total nameplate rating of 10,672 hp (7,961 kW), plus other mining equipment.

Faced with these challenges, the mine management formulated a ventilation planning method called Requirements and Resources Analysis. A requirement is defined as the air quantity or the amount of refrigeration necessary to meet the goal stated above. A resource is a tangible thing: a fan, cooler, or an airway. A properly designed system has the resources to meet the requirements. If not, work areas will be excessively warm or cold, or will suffer from high contaminant concentrations. Requirements and Resources Analysis has four basic components: 1) Establishing and justifying design parameters, 2) ascertaining present ventilation system status, 3) projecting ventilation requirements, and 4) analyzing ventilation resources alternatives.

Design parameters specify what is too hot or too cold, what concentrations of contaminants are too high, and what specific air quantities and velocities are used for specific operations. For heat, the economic temperature range for work is 80° to 85°F wet-bulb, designated as the design reject temperature. The 85° to 91°F wet-bulb increment is considered the safety-factor range where only temporary work is permitted. Above 91°F wet-bulb, ingress is for short duration only.

For dust, toxic and dangerous gases, the guidelines set forth by Mine Safety and Health Administration (MSHA) and the American Conference of Governmental Industrial Hygienists (ACGIH) are followed. For diesel equipment, a minimum 110 cfm per rated horsepower (0.07 m3/sec per kW) is specified for ramp systems and general work areas. Dead and auxiliary fans can get by with less air if diesel vehicles are only in the heading part-time. Scrams under 100 ft in length usually are well enough ventilated by convection currents and the reciprocating action of the loader.

Safety Lamp

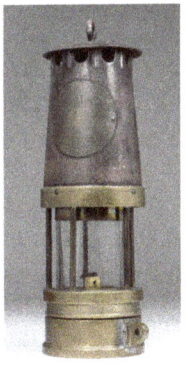

Safety lamp is a lighting device used in places, such as mines, in which there is danger from the explosion of flammable gas or dust. In the late 18th century a demand arose in England for a miner's lamp that would not ignite the gas methane (firedamp), a common hazard of English coal mines. W. Reid Clanny, an Irish physician, invented a lamp about 1813 in which the oil-fuelled flame was separated from the atmosphere by water seals; it required continual pumping for operation. In 1815 the English engineer George Stephenson invented a lamp that kept explosive gases out by pressure of the flame's exhaust and held the flame in by drawing in air at high speed. In 1815 Sir Humphry Davy invented the lamp that bears his name. Davy used a two-layer metal gauze chimney to surround and confine the flame and to conduct the heat of the flame away.

Electric hand and cap lamps were introduced in mines in the early 1900s and by the middle of the 20th century were used almost exclusively in mines. A safety device in the headpiece of the electric lamps shuts off the current if a bulb is broken. Double-filament bulbs may be used, so the light can remain on when a filament fails.

The flame of a safety lamp elongates in the presence of firedamp, but electric lamps give no warning of noxious gases or lack of oxygen. Consequently, a flame safety lamp must be kept burning within easy view of the workers, or frequent inspections must be made, using a flame lamp or other form of warning device.

Oil Lamps

Principles of Operation

Safety lamps have to address the following issues:

- Provide adequate light

- Do not trigger explosions

- Warn of a dangerous atmosphere

Fire requires three elements to burn: fuel, oxidant and heat; the triangle of fire. Remove one element of this triangle and the burning will stop. A safety lamp has to ensure that the triangle of fire is maintained inside the lamp, but cannot pass to the outside.

- Fuel - this is the easiest part of the triangle to consider, there is fuel in the form of oil inside the lamp and fuel in the form of firedamp or coal dust outside.

- Oxidant - there is an oxidant in the form of air present outside the lamp. The design of the lamp must allow the oxidant to pass into the lamp (and therefore exhaust gases to escape) or else the lamp will extinguish.

- Heat - heat can be carried by the exhaust gases, through conduction and through burning of firedamp drawn into the lamp passing back down the inlet. Control of the transfer of heat is the key to manufacturing a successful safety lamp.

In the Geordie lamp the inlet and exhausts are kept separate. Restrictions in the inlet ensure that only just enough air for combustion passes through the lamp. A tall chimney contains the spent gases above the flame. If the percentage of firedamp starts to rise, less oxygen is available in the air and combustion is diminished or extinguished. Early Geordie lamps had a simple pierced copper cap over the chimney to further restrict the flow and to ensure that the vital spent gas did not escape too quickly. Later designs used gauze for the same purpose and also as a barrier in itself. The inlet is through a number of fine tubes (early) or through a gallery (later). In the case of the gallery system air passes through a number of small holes into the gallery and through gauze to the lamp. The tubes both restrict the flow and ensure that any back flow is cooled. The flame front travels more slowly in narrow tubes (a key Stephenson observation) and allows the tubes to effectively stop such a flow.

In the Davy system a gauze surrounds the flame and extends for a distance above forming a cage. All except the very earliest Davy lamps have a double layer at the top of the cage. Rising hot gases are cooled by the gauze, the metal conducting the heat away and being itself cooled by the incoming air. There is no restriction on the air entering the lamp and so if firedamp is entrained it will burn within the lamp itself. Indeed, the lamp burns brighter in dangerous atmospheres thus acting as a warning to miners of rising firedamp levels. The Clanny configuration uses a short glass section around the flame with a gauze cylinder above it. Air is drawn in and descends just inside the glass, passing up through the flame in the centre of the lamp.

The outer casings of lamps have been made of brass or tinned steel. If a lamp bangs against a hard piece of rock a spark could be generated if iron or untinned steel were employed.

Examples of Lamps

Davy Lamp

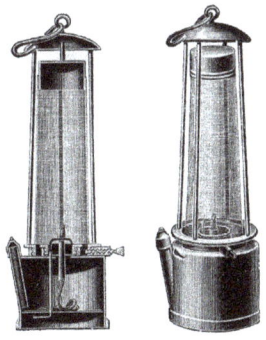

A Davy lamp

In the Davy lamp a standard oil lamp is surrounded by fine wire gauze, the top being closed by a double layer of gauze.

If firedamp is drawn into the flame it will burn more brightly and if the proportions are correct may even detonate. The flame on reaching the gauze fails to pass through and so the mine atmosphere is not ignited. However, if the flame is allowed to play on the gauze for a significant period, then it will heat up, sometimes as far as red heat. At this point it is effective, but in a dangerous state. Any further increase in temperature to white heat will ignite the external atmosphere. A sudden draught will case a localised hot spot and the flame will pass through. At a draught of between 4 and 6 feet per second the lamp becomes unsafe. At Wallsend in 1818 lamps were burning red hot (indicating significant firedamp). A boy was employed to carry hot lamps to the fresh air and bring cool lamps back. For some reason he stumbled; the gauze was damaged and the damaged lamp triggered the explosion. At Trimdon Grange a roof fall caused a sudden blast of air and the flame passed through the gauze with fatal results (69 killed).

Poor copies and ill-advised "improvements" were known, but changing dimensions either reduced the illumination or the safety. The poor light compared to either the Geordie or Clanny eventually led to the Davy being regarded as not a lamp but a scientific instrument for detecting the presence of firedamp. Some pits continued to use candles for illumination, relying on the Davy to warn men when to extinguish them.

Stephenson (Geordie) Lamp

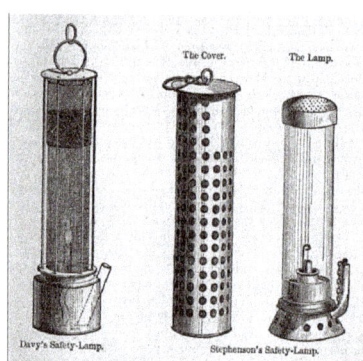

Early form of Stephenson lamp shown with a Davy lamp on the left

In the earlier Geordie lamps an oil lamp is surrounded by glass. The top of the glass has a perforated copper cap with a gauze screen above that. The glass is surrounded by a perforated metal tube to protect it.

Later versions had an annular chamber around the base of the lamp into which air entered through small (1/20") holes and from which air passed through gauze into the lamp. The glass was surrounded by gauze so that in the event of a glass breakage the Geordie became a Davy.

In a strong enough current of air enough air could be forced in through the tubes (later holes and gallery) to sustain the flame and the lamp could get red-hot. The lamp becomes unsafe in a current of from 8 to 12 feet per second, about twice that of the Davy.

Purdy Lamp

A development of the Geordie lamp was the Purdy. A galley with gauze provided the inlet, above the glass was a chimney with perforated copper cap and gauze outer. A brass tube protected the upper works, shielded them and kept them locked in position. A sprung pin locked the whole together.The pin could only be released by applying a vacuum to a captive hollow screw; not something that a nicotine starved miner could do at the coal face.

Improved Clanny Lamp

Clanny abandoned his pumps and candles and developed a safety lamp which combined features of both the Davy and Geordie. The oil lamp was surrounded by a glass chimney with no ventilation from below. Above the chimney is a gauze cylinder with a double top. Air enters from the side and spent gases exit from the top. In the presence of firedamp the flame intensifies. The flame must be kept fairly high in normal use, a small flame permits the enclosed space to fill with firedamp/air mixture and the subsequent detonation may pass through the gauze. A larger flame will keep the upper part full of burnt gas. The Clanny gives more light than the Davy and can be carried more easily in a draught. Lupton notes however it is superior in no other respect, particularly as a test instrument.

The glass on a Clanny was secured by a large diameter brass ring which could be hard to tighten securely. If a splinter occurred at the end of a crack, or indeed any other unevenness, then the seal might be compromised. Such an incident occurred at Nicholson Pit in 1856 on a lamp being used by an overman to test for firedamp. The mines inspector recommended that only Stephenson lamps were used for illumination and Davys for testing. In particular "overmen whose lamps are mostly used to detect the presence gas, should avoid such lamps".

Mueseler Lamp

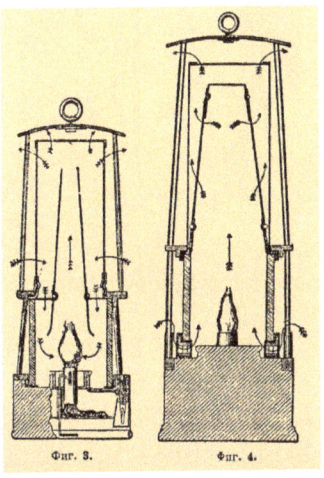

Mueseler lamp (on the left) and a derivative of the Geordie

The lamp is a modified Clanny designed by the Belgian Mathieu-Louis Mueseler. The flame is surrounded by a glass tube surmounted by a gauze capped cylinder. Air enters from the side above the glass and flows down to the flame before rising to exit at the top of the lamp. So far this is just a Clanny, but in the Mueseler a metal chimney supported on an internal gauze shelf conducts the combustion products to the top of the lamp. Some Mueseler lamps were fitted with a mechanism which locked the base of the lamp. Turning down the wick eventually released the base, but by then the flame was extinguished and therefore safe.

The lamp was patented in 1840 and in 1864 the Belgian government made this type of lamp compulsory.

In the presence of firedamp the explosive mixture is drawn through two gauzes (cylinder and shelf), burnt and then within the chimney are only burnt gases, not explosive mixture. Like a Clanny, and the Davy before it, it acts as an indicator of firedamp, burning more brightly in its presence. Later models had graduated shields by which the deputy could determine the concentration of firedamp from the heightening of the flame. Whilst the Clanny will continue to burn if laid on its side, potentially cracking the glass; the Mueseler will extinguish itself due to the stoppage of convection currents. The lamp is safe in currents up to 15 feet per second.

Marsaut Lamp

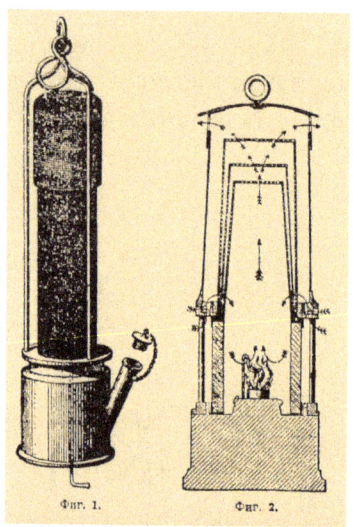

Marsaut lamp (on the right) showing a triple gauze variant

The Marsaut lamp is a Clanny with multiple gauzes. Two or three gauzes are fitted inside each other which improves the safety in a draught. Multiple gauzes will however interfere with the flow of air. The Marsaut was one of the first lamps to be fitted with a shield, in the illustration (right) the bonnet can be seen surrounding the gauzes. A shielded Marsaut lamp can resist a current of 30 feet per second.

Bainbridge Lamp

The Bainbridge is a development of the Stephenson. A tapered glass cylinder surrounds the flame, and above that the body is a brass tube. The top of the tube is closed by a horizontal gauze attached to the body of the lamp by small bars to conduct heat away. Air enters through a series of small holes drilled in the lower brass ring supporting the glass.

Landau's Lamp

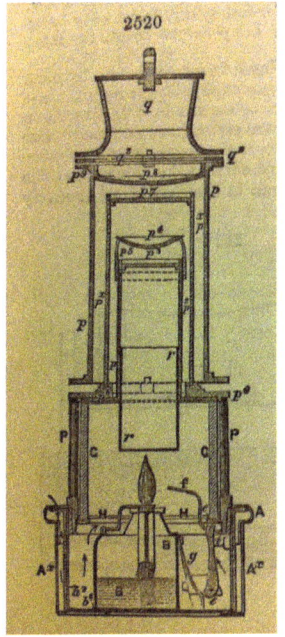

Miner's safety lamp

The lamp is in part a development of the Geordie. Air enters into a ring near the base which is protected by gauze or perforated plate. The air passes down the side of the lamp passing through a series of gauze covered holes and enters the base through another yet another series of gauze covered holes. Any attempt to unscrew the base causes the lever (shown at f in the illustration) to extinguish the flame. The gauze covered holes and passageways restrict the flow to that required for combustion, so if any part of the oxygen is replaced by firedamp, then the flame is extinguished for want of oxidant.

The upper portion of the lamp uses a chimney like Mueseler and Morgan lamps. The rising gases pass up the chimney and through a gauze. At the top of the chimney a dished reflector diverts the gases out sideways through a number of holes in the chimney. The gases then start to rise up the intermediate chimney before exiting through another gauze. Gas finally passes down between the outermost chimney and the intermediate chimney, exiting a little above the glass. The outer chimney is therefore effectively a shield.

Yates' Lamp

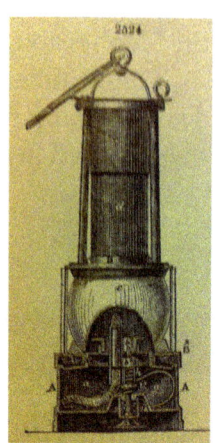

Miner's safety lamp

The Yates lamp is a development of the Clanny. Air enters through the lower part of the gauze top and leaves through the upper part; there is no chimney. The lower glass part of the lamp has seen some development however. It is replaced by a silvered reflector having a strong lens or bull's-eye in it to allow the light out. The result was a claimed 20 fold improvement in lighting over the Davy. Yates claimed "the temptation to expose the flame to obtain more light is removed".

The base also contained an interlocking mechanism to ensure that the wick was lowered and the lamp extinguished by any attempt to open it.

The lamp was "much more expensive than the forms of lamp now in general use, but Mr, Yates states that the saving of oil effected by its use will in one year pay the additional cost".

Evan–thomas

The Evan–Thomas lamp is similar to a shielded Clanny, but there is a brass cylinder outside the gauze above the glass. It resists draughts well but the flame is dull.

Morgan

The Morgan is a cross between the Mueseler and the Marsaut. It is a shielded lamp with a series of disks at the top to allow spent fumes out and a series of holes lower down the shield to allow air in. There is an inner and outer shield so that air cannot blow directly on the gauze but must first find its way through a slim chamber. There are multiple gauzes, like the Mersaut, and there is an internal chimney like the Mueseler. There is no "shelf" supporting the chimney, instead it hangs from an inverted cone of gauze.

The Morgan will resist air up to 53 feet per second and is "sufficiently safe for every practical purpose".

Clifford

The Clifford also has a double shield, but with a plain flat top. The chimney is quite narrow with gauze covering the top. The bottom of the chimney has a glass bell covering the flame. The chimney is supported on a gauze shelf. Air enters through the lower part of the outer shield, through the passage and into the lamp through the inner shield. It is drawn down through the gauze then passes the flame and ascends the chimney. At the top it leaves through gauze and the top of the double shield. The inner chimney is made of copper coated with a fusible metal. If the lamp gets too hot the metal melts and closes up the air holes, extinguishing the lamp.

The lamp has been tested and according to Lupton "successfully resisted every effort to explode it up to a velocity of more than 100 feet per second".

Electric Lamps

It was not until tungsten filaments replaced carbon that a portable electric light became a reality. An early pioneer was Joseph Swan who exhibited his first lamp in Newcastle upon Tyne in 1881 and improved ones in subsequent years. The Royal Commission on Accidents in Mines set up in 1881 carried out extensive tests of all types of lamps and the final report in 1886 noted that there had been good progress made in producing electric lamps giving a light superior to that of oil lamps and expected economic and efficient lamps to become available soon. This turned out not to be the case and progress was slow in attaining reliability and economy. The Sussmann lamp was introduced into Britain in 1893 and following trials at Murton Colliery in Durham it became a widely used electric lamp with 3000 or so reported by the company in use in 1900 However, by 1910 there were only 2055 electric lamps of all types in use - about 0.25% of all safety lamps. In 1911, an anonymous colliery owner, through the British government, offered a prize of £1000 (equivalent to £93,459 in 2016) for the best lamp to specified requirements. There were 195 entries. It was won by a German engineer with the CEAG lamp,which was hand-held and delivered twice the illumination of oil lamps, with a battery life of 16 hours. Awards were made to 8 other lamps that met the judges' criteria. Clearly this stimulated development and over the next few years there was a marked increase in the use of electric lamps, especially the CEAG, Gray-Sussmann and Oldham, so by 1922 there were 294593 in use in Britain.

In 1913, Thomas Edison won the Ratheman medal for inventing a lightweight storage battery that could be carried on the back, powering a parabolic reflector that could be mounted on the miner's helmet. After extensive testing, 70,000 robust designs were in use in the US by 1916,.

Early electric lamps in Britain were hand held as miners were used to this and helmet lamps became common much later than in countries like the USA where helmet (cap) lamps had been the norm.

Nowadays, safety lamps are mainly electric, and traditionally mounted on miners' helmets (such as the wheat lamp) or the Oldham headlamp, sealed to prevent gas penetrating the casing and being ignited by electrical sparks.

Although its use as a light source was superseded by electric lighting, the flame safety lamp has continued to be used in mines to detect methane and blackdamp, although many modern mines now also use sophisticated electronic gas detectors for this purpose.

As a new light source, LED has many advantages for safety lamps, including longer illumination times and reduced energy requirements. Combined with new battery technologies, such as the lithium battery, it gives much better performance in safety lamp applications. It is replacing conventional safety lamps.

The Office of Mine Safety and Health (OMSHR), a part of the National Institute for Occupational Safety and Health (NIOSH) (itself part of Centers for Disease Control and Prevention) in the United States has been investigating the benefits of LED headlamps. A problem in mining is that the average age is increasing: 43.3 years in 2013 (in the USA) and as a person ages vision degenerates. LED technology is physically robust compared to a filament light bulb, and has a longer life: 50,000 hours compared to 1,000 – 3,000. Extended life reduces light maintenance and failures; according to OMSHR an average of 28 accidents per year occur in US mines involving lighting. NIOSH has sponsored the development of cap lamp systems which they claim improve the "ability of older subjects to detect moving hazards by 15% and trip hazards by 23.7%, and discomfort glare was reduced by 45%". Conventional lights are strongly focussed in a beam, NIOSH LED lamps are designed to produce a wider more diffuse beam which is claimed to improve the perception of objects by 79.5%.

Flame Safety Lamp

Flame Safet lamp used for detection of methane gas in mines. The colour and height of the flame in a flame safety lamp can indicate the presence of firedamp and maybe estimate the percentage in the air. The flame safety lamp can be divided into three main sections – lower, middle and upper sections. There are various components in all the three sections. Table lists the different components found in different section along with their functional details.

Table: Components of flame safety lamp in different sections along with their functional details

S. No.	Section	Name of the components	Functions /details
1	Lower	Fuel vessel, burner, wick assembly, a screw, locking arrangement, and re-lighting device(if available)	Locking arrangement is generally of magnetic type. It can be opened only in the lamp room. Thus, once it is locked, it cannot be opened in the underground.
2	Middle	Two sets of rings interconnected by five steel rods, glass (may be one or two set-inner and outer) around the flame, chimney, two asbestos washer	Ring and rod combination serves to protect the glass. Rings have holes to allow air in/out. Chimney is introduced to increase the illumination produced. Asbestos washer is used to make the assembly air tight.
3	Upper	A bonet, two steel wire gauzes(of 20 or 28 mesh size), handle / hook	It is the wire gauze which is responsible for the working of flame safety lamp and not allowing the ignition of methane, even if present in the general body of air. Handle serves the purpose of holding the lamp while detecting methane.

Table gives only a brief idea about the construction of the flame safety lamp.

Working Principle of Flame Safety Lamp

It is very interesting to know that wire gauze does not let methane to ignite even if there is methane in the general body of air. The ignition of methane remains limited only to the interior of the lamp. For a lay man it would be like a magic. To understand the working principle of the flame safety lamp.

Figure shows a Bunsen burner and wire gauze. The distance between the wire gauze and the mouth of the Bunsen burner is kept to be around 30 mm. The gas is turned on and then by using match box or suitable device it is ignited below the gauze. It is observed that the flame of the burning gas is restricted below the gauze only and is not able to cross the gauze wire until the gauze wire is red hot. Once the wire is red hot, the flame can be seen above as well as below the gauze wire.

Figure shows the same arrangement, but flame of the gas is observed above the gauze wire. This is because the gas was ignited above the gauze wire. Here also, the flame appears below as well as above the gauze wire only after the gauze wire is red hot.

It is interesting to know that, if gauze wire is lowered to the mouth of the Bunsen burner or moved upward, the flame gets extinguished, even if gas supply is turned on. However, this is possible only when it is done before the gauze wire becomes red hot.

The gauze wire does not allow flame to pass through it until it becomes red–hot. The reason behind such behavior of the gauze wire is described below.

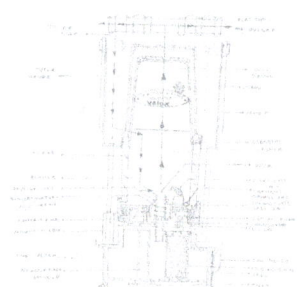 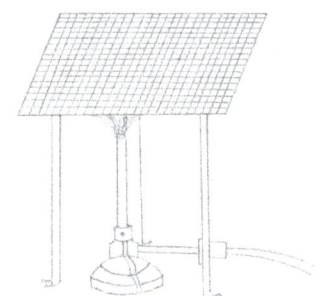 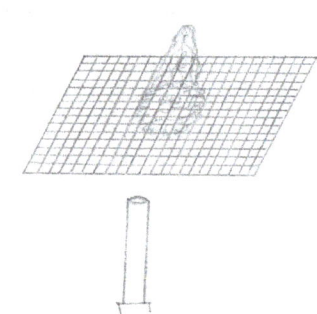

Figure: Various components of   Figure: Principle of wire gauze   Figure: Principle of wire
flame safety lamp   gauze

Gauze wire is made up of metals like copper or iron. It may be made up of steel also. These metals allow the gas to pass through it. But, they are very good conductor of heat also. They conduct heat so quickly that the temperature above the gauze wire is below the ignition temperature of the gas. But, once the wire becomes red hot, it is not able to conduct the heat, and the temperature above it crosses the ignition temperature of the gas. Thus, gas gets ignited and flame appears on both the sides of the gauze wire.

A dismantled flame safety lamp is assembled by bringing the upper section and the middle section together. This assembly is then screwed onto the lower section and it gets magnetically locked. Like wire gauze, even locking arrangement in a flame safety lamp is one of the important safety device. A properly assembled safety lamp will not produce any sound due to loose components, if it is shaken by hand. The locking arrangement consists of a spring loaded steel bolt housed in a tubular body which is fitted and soldered with the bottom flange of the middle section. The magnetic lock bolt passes through the collar into the notches on the oil vessel. When the middle and top sections are fitted on the oil vessel by screwing, the lock bolt prevents their unscrewing by the ratchet construction at the top end of the oil vessel. When we want to unlock, the top of magnetic locking device is placed below the pole of magnet unlocker in the lamp cabin. The lock bolt is pulled by the magnet and the base of the lamp can then be unscrewed. The magnetic locking arrangement is so designed that ordinary magnet cannot unlock the lamp. Figure: (a), (b) and (c) clearly show the lock bolt, ratchet arrangement at the top end of the oil vessel and magnetic unlocker respectively.

Figure (a): Lock bolt in a Flame safety Lamp

Figure (b): Ratchet arrangement at the top end of the oil vessel

Figure (c): Magnetic unlocker

Approval by DGMS

To get approved by DGMS, a flame safety lamp has to pass through several tests. Also, the safety lamp has to conform to ISI specifications No. IS 7577 or 1975. Some of the tests are:-

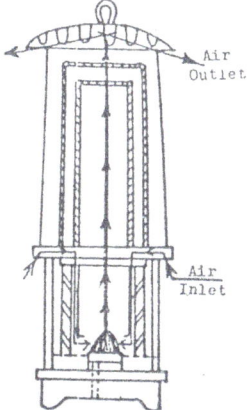

Figure: Mid-feed flame safety lamp

- Drop test :- The lamp has to be dropped for 5 times from a height of 1 m (measured from the bottom of lamp to a wooden board) on to a hard wooden board. The board should be 30 mm thick and should be laid on concrete. The test is passed by lamp only if no damage to any component of the lamp is observed.

- Test for air tightness :- After assembling the different sections of the lamp, it is lit and held before a compressed air jet of velocity not less than 6 m/s. The test is said to be passed if flame is not extinguished. Further, no undue flickering should be observed.

- Performance test :- This is decided on the basis of length of the flame produced as well as the condition. By condition of flame, it is meant for intensity, colour and its sensitivity to a slight change in the concentration of the methane. Table lists the length of flame and condition of flame for the lamp to pass through this test at different concentrations of methane in the air.

Table: Performance test on flame safety lamp

Gas %	Minimum length of blue flame (mm)	Condition of blue flame
0.0	-	Near the top of the standard flame a slight cobalt-blue lined orange-yellow flame is seen
1.0	7.0	Blue flame is not observed because the flame color is light and hence it is difficult to measure the length
1.5	8.0	Lower part of flame turns blue
2.0	9.0	Blue flame is distinctly visible except at the top
2.5	10.0	Blue color becomes more visible, but still top is invisible
3.0	11.5	Top of flame is still invisible and blue flame is seen clearer
3.5	14.5	Blue flame is clearly visible
4.0	20.0	Blue flame is extremely clear, highly sensitive to slight change in methane concentration

Please have a look at figure for better understanding of Table.

Some of the more specifications are:-

- Rivet locking arrangement shall not be of lead type.

- Thickness of glass cylinder shall be 4-5 mm.

- Total heat radiation area of the gauze should not be less than 155 cm².

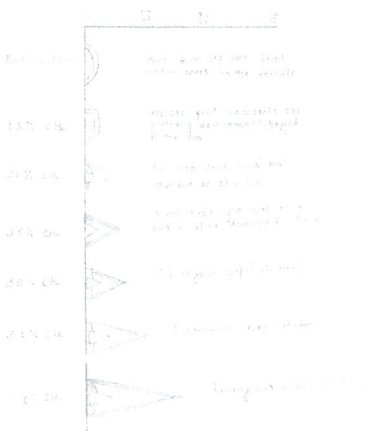

Figure: Variation in shape and height of gas cap with methane percentage

Geordie Lamp

The Geordie lamp was a safety lamp for use in inflammable atmospheres, invented by George Stephenson in 1815 as a miner's lamp to prevent explosions due to firedamp in coal mines.

Origin

In 1815, Stephenson was the engine-wright at the Killingworth Colliery in Northumberland and had been experimenting for several years with candles close to firedamp emissions in the mine. In August he ordered an oil lamp which was delivered on 21 October and tested by him in the mine in the presence of explosive gases. He improved this over several weeks with the addition of capillary tubes at the base so that it gave more light and tried new versions on 4 and 30 November. This was presented to the Literary and Philosophical Society of Newcastle upon Tyne on 5 December 1815.

Although controversy arose between Stephenson's design and the Davy lamp , Stephenson's original design worked on significantly different principles from Davy's final design. If the lamp were sealed except for a restricted air ingress (and a suitably sized chimney) then the presence of dangerous amounts of firedamp in the incoming air would (by its combustion) reduce the oxygen concentration inside the lamp so much that the flame would be extinguished. (Stephenson had convinced himself of the validity of this approach by his experiments with candles near lit blowers: as lit candles were placed upwind of the blower the blower flame grew duller; with enough upwind candles the blower flame went out.) To guard against the possibility of a flame travelling back through the incoming gases (an explosive backblast), air ingress was by a number of small-bore orifices through which the air flowed at a velocity higher than the velocity of the flame in a mixture of firedamp (mostly methane) and air. The body of the lamp was lengthened to give the flame a greater convective draw, and thus allow a greater inlet flow restriction and make the lamp less sensitive to air currents. Davy had originally attempted a safety lamp on similar principles, before preferring to enclose the flame inside a brass gauze cylinder; he had publicly identified the importance of allowing the restricted airflow in through small orifices (in which the flame velocity is lower) before Stephenson had, and he and his adherents remained convinced that Stephenson had not made this discovery independently.

One advantage of Stephenson's design over Davy's was that if the proportion of fire-damp became too high, his lamp would be extinguished, whereas Davy's lamp could become dangerously hot. This was illustrated in the Oaks colliery at Barnsley on 20 August 1857 where both types of lamp were in use.

Stephenson's design used glass to surround the flame, which cut out less of the light than Davy's, where the gauze surrounded it. But this also posed the danger of breakage in the harsh conditions of mineworking, a problem which was not resolved until the invention of safety glass. Stephenson tried several different designs in early years and later adopted Davy's gauze in preference to the tubes and it was this revised design that was used for most of the 19th century as the Geordie lamp.

The Geordie lamp continued to be used in the north-east of England through most of the 19th century, until the introduction of electric lighting.

Wheat Lamp

A wheat lamp is a type of incandescent light designed for use in underground mining, named for inventor Grant Wheat and manufactured by Koehler Lighting Products in Wilkes-Barre, Pennsylvania, United States, a region known for extensive mining activity.

A safety lamp designed for use in potentially hazardous atmospheres such as firedamp and coal dust, the lamp is mounted on the front of the miner's helmet and powered by a wet cell battery worn on the miner's belt. The average wheat lamp uses a 3-5 watt bulb which will typically operate for 5 to 16 hours depending on the amp-hour capacity of the battery and the current draw of the bulb being used.

A grain of wheat lamp is an unrelated, very small incandescent lamp used in medical and optical instruments, as well as for illuminating miniature railroad and similar models.

Cap Lamp

The cap lamp or more accurately 'Electric Safety Lamp' is powered by a battery carried on the miner's belt and a cable running up to the lamp. The lamp may be slotted onto the front of the miner's helmet by means of a clip or can be detached and used as a flashlight for performing tasks underground. The sight of groups of miners walking through tunnels or being transported to the surface in the 'cage' illuminated by the light of their cap lamps.

Cap lamps are issued from a lamp-room located on the pit-top where units are charged and tested prior to the next shift. The design has found other uses above ground, notably, applications in the fire and rescue service.

Personal Protective Equipment in Mining

Head Protection

In most countries miners must be provided with, and must wear, safety caps or hats which are approved in the jurisdiction in which the mine operates. Hats differ from caps in that they have a full brim rather than just a front peak. This has the advantage of shedding water in mines which are very wet. It does, however, preclude the incorporation of side slots for mounting of hearing protection, flashlights and face shields for welding, cutting, grinding, chipping and scaling or other accessories. Hats represent a very small percentage of the head protection worn in mines.

The cap or hat would in most cases be equipped with a lamp bracket and cord holder to permit mounting of a miner's cap lamp.

The traditional miner's cap has a very low profile which significantly reduces the propensity for the miner to bump his or her head in low seam coal mines. However, in mines where head room is adequate the low profile serves no useful purpose. Furthermore, it is achieved by reducing the clearance between the crown of the cap and the wearer's skull so that these types of cap rarely meet the top impact standards for industrial head protection. In jurisdictions where the standards are enforced, the traditional miner's cap is giving way to conventional industrial head protection.

Standards for industrial head protection have changed very little since the 1960s. However, in the 1990s, the boom in recreational head protection, such as hockey helmets, cycle helmets and so on, has highlighted what are perceived to be inadequacies in industrial head protection, most notably lack of lateral impact protection and lack of retention capabilities in the event of an impact. Thus, there has been pressure to upgrade the standards for industrial head protection and in some jurisdictions this has already happened. Safety caps with foam liners and, possibly, ratchet suspensions and/or chin straps are now appearing in the industrial marketplace. They have not been widely accepted by users because of the higher cost and weight and their lesser comfort. However, as the new standards become more widely entrenched in labour legislation the new style of cap is likely to appear in the mining industry.

Eye and Face Protection

Most mining operations around the world have compulsory eye protection programmes which require the miner to wear safety spectacles, goggles, faceshields or a full facepiece respirator, depending on the operations being performed and the combination of hazards to which the miner is exposed. For the majority of mining operations, safety spectacles with side shields provide suitable protection. The dust and dirt in many mining environments, most notably hard-rock mining, can be highly abrasive. This causes scratching and rapid wear of safety glasses with plastic (polycarbonate) lenses. For this reason, many mines still permit the use of glass lenses, even though they do not provide the resistance to impact and shattering offered by polycarbonates, and even though they may not meet the prevailing standard for protective eye wear in the particular jurisdiction. Progress continues to be made in both anti-fog treatments and surface hardening treatments for plastic lenses. Those treatments which change the molecular structure of the lens surface rather than simply applying a film or coating are typically more effective and longer lasting and have the potential to replace glass as the lens material of choice for abrasive mining environments.

Goggles are not worn frequently below ground unless the particular operation poses a danger of chemical splash.

A faceshield may be worn where the miner requires full-face protection from weld spatter, grinding residues or other large flying particles which could be produced by cutting, chipping or scaling. The faceshield may be of a specialized nature, as in welding, or may be clear acrylic or polycarbonate. Although faceshields can be equipped with their own head harness, in mining they will normally be mounted in the accessory slots in the miner's safety cap. Faceshields are designed so that they can be quickly and easily hinged upwards for observation of the work and down over the face for protection when performing the work.

A full facepiece respirator may be worn for face protection when there is also a requirement for respiratory protection against a substance which is irritating to the eyes. Such operations are more often encountered in the above ground mine processing than in the below ground mining operation itself.

Respiratory Protection

The most commonly needed respiratory protection in mining operations is dust protection. Coal dust as well as most other ambient dusts can be effectively filtered using an inexpensive quarter facepiece dust mask. The type which uses an elastomer nose/mouth cover and replaceable filters is effective. The moulded throw-away fibre-cup type respirator is not effective.

Welding, flame cutting, use of solvents, handling of fuels, blasting and other operations can produce air-borne contaminants that require the use of twin cartridge respirators to remove combinations of dust, mists, fumes, organic vapours and acid gases. In these cases, the need for protection for the miner will be indicated by measurement of the contaminants, usually performed locally, using detector tubes or portable instruments. The appropriate respirator is worn until the mine ventilation system has cleared the contaminant or reduced it to levels that are acceptable.

Certain types of particulates encountered in mines, such as asbestos fibres found in asbestos mines, coal fines produced in longwall mining and radionuclides found in uranium mining, may require the use of a positive pressure respirator equipped with a high-efficiency particulate absolute (HEPA) filter. Powered air-purifying respirators (PAPRs) which supply the filtered air to a hood, tight-fitting facepiece or integrated helmet facepiece assembly meet this requirement.

Hearing Protection

Underground vehicles, machinery and power tools generate high ambient noise levels which can create long-term damage to human hearing. Protection is normally provided by ear muff type protectors which are slot-mounted on the miner's cap. Supplementary protection can be provided by wearing closed cell foam ear plugs in conjunction with the ear muffs. Ear plugs, either of the disposable foam cell variety or the reusable elastomeric variety, may be used on their own, either because of preference or because the accessory slot is being used to carry a face shield or other accessory.

Skin Protection

Certain mining operations may cause skin irritation. Work gloves are worn whenever possible in such operations and barrier creams are provided for additional protection, particularly when the gloves cannot be worn.

Foot Protection

The mining work boot may be of either leather or rubber construction, depending on whether the mine is dry or wet. Minimum protective requirements for the boot include a full puncture-proof sole with a composite outer layer to prevent slipping, a steel toe-cap and a metatarsal guard. Although these fundamental requirements have not changed in many years, advances have been made towards meeting them in a boot that is far less cumbersome and far more comfortable than the boots of several years ago. For example, metatarsal guards are now available in moulded fibre, replacing the steel hoops and saddles that were once common. They provide equivalent protection with less weight and less risk of tripping. The lasts (foot forms) have become more anatomically correct and energy absorbing mid-soles, full moisture barriers and modern insulating materials have made their way from the sports/recreation footwear market into the mining boot.

Clothing

Ordinary cotton coveralls or treated flame-resistant cotton coveralls are the normal workwear in mines. Strips of reflective material are usually added to make the miner more visible to drivers of moving underground vehicles. Miners working with jumbo drills or other heavy equipment may also wear rain suits over their coveralls to protect against cutting fluid, hydraulic oil and lubricating oils, which can spray or leak from the equipment.

Work gloves are worn for hand protection. A general purpose work glove would be constructed of cotton canvas reinforced with leather. Other types and styles of glove would be used for special job functions.

Belts and Harnesses

In most jurisdictions, the miners belt is no longer considered suitable or approved for fall protection. A webbing or leather belt is still used, however, with or without suspenders and with or without a lumbar support to carry the lamp battery as well as a filter self-rescuer or self-contained (oxygen generating) self-rescuer, if required.

A full body harness with D-ring attachment between the shoulder blades is now the only recommended device for protecting miners against falls. The harness should be worn with a suitable lanyard and shock absorbing device by miners working in shafts, over crushers or near open sump or pits. Additional D-rings may be added to a harness or a miner's belt for work positioning or to restrict movement within safe limits.

Protection from Heat and Cold

In open-pit mines in cold climates, miners will have winter clothing including thermal socks, underwear and gloves, wind resistant pants or over-pants, a lined parka with hood and a winter liner to wear with the safety cap.

In underground mines, heat is more of a problem than cold. Ambient temperatures may be high because of the depth of the mine below ground or because it is located in a hot climate. Protection from heat stress and potential heat stroke can be provided by special garments or undergarments which can accommodate frozen gel packs or which are constructed with a network of cooling tubes to circulate cooling fluids over the surface of the body and then through an external heat exchanger. In situations where the rock itself is hot, heat resistant gloves, socks and boots are worn. Drinking water or, preferably, drinking water with added electrolytes must be available and must be consumed to replace lost body fluids.

Other Protective Equipment

Depending on local regulations and the type of mine, miners may be required to carry a self-rescue device. This is a respiratory protection device which will help the miner to escape from the mine in the event of a mine fire or explosion that renders the atmosphere unbreathable because of carbon monoxide, smoke and other toxic contaminants. The self-rescuer may be a filtration type device with a catalyst for carbon monoxide conversion or it may be a self-contained self-rescuer, i.e., a closed-cycle breathing apparatus which chemically regenerates oxygen from exhaled breath.

Portable instruments (including detector tubes and detector tube pumps) for the detection and measurement of toxic and combustible gases are not carried routinely by all miners, but are used by mine safety officers or other designated personnel in accordance with standard operating procedures to test mine atmospheres periodically or before entry.

Improving the ability to communicate with personnel in underground mining operations is proving to have enormous safety benefits and two-way communication systems, personal pagers and personnel locating devices are finding their way into modern mining operations.

Self-contained Self-Rescue Device

A self-contained self-rescue unit, or Air Pack, is a portable oxygen source for providing breathable air in emergency situations. Such as if the oxygen level in the surrounding atmosphere would unexpectedly drop or become contaminated with toxic gases. A SCSR is a closed-circuit breathing device intended for one person and usually provides one hour of breathable air. The oxygen is provided with a chemical oxygen generator or a compressed oxygen cylinder and a carbon monoxide absorber. They are most commonly used by personnel in the mining industry, especially in coal mines. Also in other fields where personnel are working in an environment that could be sealed of and oxygen levels can drop due to fires, explosions or cave-ins, such as tunneling. The purpose of an SCSR is to provide breathable air in emergency situations, long enough to facilitate escape from mines after a fire or explosion. Usually the goal is to reach a safe bunker and wait for rescue.

Mine Rescue

The mining industry is among the top ten industries nationwide with high occupational injury and fatality rates, and mine rescue operations are a relatively high risk activity in underground coal mining. Mine rescue team members must be prepared to respond when an emergency occurs and take the necessary precautions required to ensure worker safety. It is vital that members of the teams have the capability and proper protection to immediately respond to a wide range of hazardous situations. Their ensembles need to be able to protect them from hazards that they may encounter. In addition, mine rescue team members must know the limitations of their personal protective ensembles.

Mine Safety and Health Administration (MSHA) defines "mine rescue" as "the practiced response to a mine emergency situation that endangers life, property, and the continued operation of the mine". The primary objective of mine rescue is described as preventing loss of life, and the secondary objective is the safe recovery of the mine and its return to normal production. In its earliest days, mine rescue was an unsystematic effort. Rescue "parties" were groups of miners and other volunteers who happened to be at the mine site at the time of the disaster. These groups had no training, no equipment, and no reliable breathing gear; and frequently, their names were added to the list of those who died in the disasters.

Right after its establishment in 1910, the United States Bureau of Mines (USBM) undertook a program of obtaining railroad passenger cars and modifying them into mobile stations for mine rescue and first aid training. These cars were equipped with breathing apparatuses and carried a crew of six men, each trained for a specific duty in regards to mine rescue and first aid.

While trying to save the lives of others, mine rescue team members have been injured and killed. Since 1900, 11,719 underground coal mine workers died in 509 U.S. underground coal mining disaster incidents, with most disasters resulting from explosions. The history of anthracite coal mining in Pennsylvania was marked by an alarming increase in the number of fatalities in the late 1800s. One hundred and eight miners and two mine rescuers were killed in 1869 at the Avondale Mine in Plymouth, Luzerne County, PA when a surface fire blocked the exit of the mine. After increasing each year, the number of occupational coal mining fatalities in underground coal mines in the U.S. surpassed 500 by 1896. Figure highlights coal mining disaster incidents and the fatalities between 1900 and 2010. As a result of these fatalities, the first formal mine rescue teams were organized and trained in the 1900s.

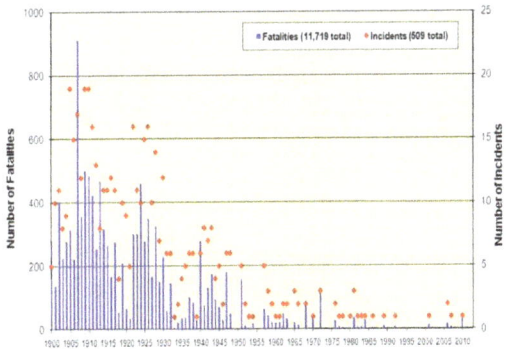

Coal Mining Disaster(*) Incidents and Fatalities (*): A mining disaster is an incident with 5 or more fatalities

Also, since the Avondale Mine Disaster in PA in 1869, 125 rescue workers were killed during the rescue efforts (figure below). The most common events of the incidents were, explosion, mine fire, inundation (the sudden inrush of water or toxic gases from old workings), seismic jolt, and mine collapse. It should be noted that these rescuers were not all members of formal mine rescue teams. Many were other miners who happened to be at the mine or in the area and responded without donning any mine rescue ensemble.

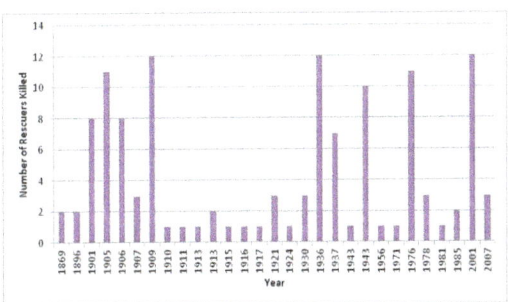

Mine Rescuers Killed at the Mine Accidents (**)

In 2006, there were three major underground coal mining accidents: Sago, Aracoma Alma and Darby Mine # 1. After three mine disasters in five months, the Miner Improvement and New Emergency Response (MINER) Act was signed on June 15, 2006 to improve accident preparedness and response. The other requirements of the MINER Act were in regards to the development of written emergency response plan, use of equipment and technology, additional mine rescue team training requirements, teams' response time, civil and criminal penalties, establishment of a competitive grant program for new mine safety technology, and an interagency working group to provide a formal means of sharing non-classified technology that would have applicability to mine safety. The MINER Act of 2006 introduced a significant change in mine rescue. However no information with regard to the minimum level personal protective equipment (PPE) required for mine rescue operations was included.

The most recent tragedy where mine rescue team members were killed, happened on August 6, 2007 when a catastrophic coal outburst accident occurred at the Crandall Canyon Mine, in Emery County, Utah. Two mine rescue team members employed by the coal company and one MSHA mine rescue team member died due to injuries received during the roof collapse. Six additional mine rescue team members, including one MSHA member, were also injured. Underground rescue efforts were suspended following these fatalities.

There is very limited information available in regard to the ensembles worn by the mine rescue teams during the mine disasters. However, it can be stated that in general, the ensembles used during the incidents include a typical mining coverall which is made of cotton or cotton/polyester blends and not fire resistant, helmet, cap lamp, boots, kneepads, facemask, breathing apparatus, belt, and suspenders. During some of the mine disasters, such as explosions, it may be extremely difficult to protect mine rescue team members and prevent the injuries and/or fatalities through the use of a more encompassing ensemble or any kind of PPE; however, in other cases such as fire fighting, providing first aid, and recovering, the injuries or fatalities can be decreased by using ensembles designed to reduce exposures or the consequences of the exposures. These ensembles should also be ergonomically designed and aid mine rescue teams in their tasks. Thus, it can be stated that there is a room to improve the safety, health and performance of mine rescue team members by specifying the mine rescue ensemble elements and identifying the minimum performance and design characteristics of the mine rescue ensembles.

Today's mine rescue efforts are highly organized and very cautiously managed operations carried out by groups of trained and skilled individuals who work together as a team. Regulations require all underground mines to have fully-trained and equipped professional mine rescue teams available in the event of a mine emergency. Currently, there are 217 underground coal mine rescue teams with a total of 1888 members in the U.S.

Mine Rescue Teams and Their Tasks

Mine rescue and recovery involves a wide variety of tasks. The way that the mine rescue teams respond varies according to the type of mine emergency and the type of the mine being entered. Conditions within the mine also determine what the team will be required to do. MSHA defines some of the tasks that may be required during an actual emergency by mine rescue teams as:

- Exploring the affected area of the mine.

- Searching for and rescuing survivors.

- Performing first aid.

- Determining the extent of damage.

- Determining gas conditions.

- Mapping the team's findings.

- Locating and fighting fires.

- Building temporary and/or permanent stoppings/bulkheads.

- Erecting seals in a fire area.

- Clearing debris, pumping water, and installing or erecting temporary roof supports.

A mine rescue team for underground coal mines consists of a minimum of five members, plus one alternate, who are fully qualified, trained, and equipped for providing emergency mine rescue response. The six team positions usually include:

- Captain who leads the team,

- Gas person who backs up the captain and checks for the presence of gas,

- Map person who maps locations of conditions in the mine and actions taken by the team,

- Stretcher person who pulls the stretcher,

- Tail person or co-captain who receives orders from the fresh air base briefing officer and relays information from inside the mine to the fresh air base, and

- Briefing officer who remains at the fresh air base and directs the teams according to Command Center order, and also informs the Command Center of mine conditions found during exploration.

Mine rescue team members.

Prior to serving on a mine rescue team, each member of a coal mine rescue team must complete, at a minimum, an initial 20-hour course of instruction as prescribed by MSHA, in the use, care, and maintenance of the type of breathing apparatus which will be used by the mine rescue team. Upon completion of the initial training, all team members are required to receive at least 96 hours of refresher training annually. This refresher training may include: a written test, bench testing of the breathing apparatus, first aid, fire fighting, locating miners, smoke training, and proper techniques for evaluating for noxious gases, mine mapping, ventilation controls, and proper techniques for examining the overall conditions of the mine. The type of clothing and equipment used by the team members do not differ by the member's role or the type of the activity (providing the first aid, fire fighting, or exploration.

Mine Rescue Teams

Threat Detection

Some of the dangers encountered in mines may include toxic gases or low oxygen levels, rotten timbering or no ground support at all, unseen dry-rot in bulkheads, invisible vertical shafts to lower levels, poisonous insects or snakes, frightened animals, and old explosives.

When rescue teams are called out in irrespirable atmospheres (toxic gasses or lack of oxygen) and under difficult climatic conditions, the following dangers may arise:

- Carbon monoxide and/or carbon dioxide poisoning

- Lack of oxygen

- Circulatory control failures

- Heat build up

Human error, self-overestimation, lacking physical conditions, nervous stress as well as leaking breathing tube connections and faulty equipment or accessories may lead to accidents.

From the literature review and meetings conducted with the mining personnel from extensive locations of U.S., it was determined that the main events for mine rescue fatalities and injuries of the 30 previous coal mine disasters include, coal bump/ bounce (e.g., Crandall Canyon mine disaster), explosions (e.g., Scotia mine disaster), heat stress, slips and falls, roof falls, rib rolls (a slab of coal from a left over block of coal comes loose), asphyxiation (suffocation), burns from fire, overcome in a rescue, drowning during fire fighting, and overcome by carbon monoxide. Sometimes rescue effort without PPE or adequate gas test equipment resulted in the injuries or fatalities. During some of the mine disasters, such as explosions, it may be extremely difficult to protect mine rescue team members through the use of any kind of PPE, however, in other cases such as fire fighting and providing first aid, the injuries or fatalities may be reduced by evaluating the hazardous conditions of the events and using ensembles designed to meet the needs.

It is crucial to understand the operating environment/hazards and duties to investigate the requirements for PPE for mine rescuers. According to a study report presented to USBM, environmental conditions for mine rescue operations can be summarized in the following categories:

- Toxic Gases, Smoke and Particulate Matter: Methane, carbon monoxide, carbon dioxide, hydrogen, nitrogen oxide, sulphur dioxide, ethane, propane, butane, smoke, and other toxic and irritating material require the use of adequate respiratory protection.

- Temperature: The temperature in an underground coal mine varies with condition and location. Field study data show that a mine rescue team may operate in an environment ranging between 50°F and 150°F and on occasion may be exposed to even higher temperatures. The exposure time to the high temperature is usually no more than a few minutes because of the limits of human endurance.

- Heat: There are three modes of heat transfer: conduction, convection, and radiation. In mine rescue situations, the contact temperature can be as high as 1000°F-1200°F (conduction) and hot gas temperatures can range from 100°F-1500°F (convection). Flames are the greatest source of radiant energy but other materials may radiate too. At fire scenes, where direct contact is not made with a hot object, the heat load is comprised of both radiant and convective fractions, with convection being the small portion of the total heat loads.

- Flame: The rescue and recovery team is infrequently in direct contact with flame. Whenever contact is made with flame, it is usually the result of a falling ember and only lasts a few seconds. However, rescue teams called upon to fight fire will be directly exposed to flames.

- Water: The primary problems associated with water arise when the team gets wet, possibly soaked, all the way through their undergarments. The clothing becomes uncomfortable, and the weight of the water absorbed contributes substantially to fatigue. Also, a wet garment may result in steam burns, if the mine rescue team member suddenly comes in contact with a heat source.

- Bloodborne Pathogens: During the recovery of injured miners and providing first-aid, teams may be exposed to bloodborne pathogens from blood and body fluids.

Apparatus used in Mine Rescue

The mine rescue ensemble is defined as the integrated elements of the rescue team's personal protection system. The function of the mine rescue ensemble is to improve the wearer's protection against hazards such as heat, flame, toxic gases, smoke, penetration, impact and water. The elements include:

- Protective Garments and Equipment for Body Protection - helmet and hood for head protection, ear protection (rarely used), coverall or pant/jacket for torso and limb protection, gloves for hand and wrist protection, kneepads for knee protection, and boots for foot and ankle protection.

- Respiratory Protection - Closed Circuit Self Contained Breathing Apparatus with a full facepiece (SCBA).

- Lighting System - cap lamp with a cord and battery, or cordless.

- Communication Systems – portable radios or hard-wired communication systems, etc.

- Navigation Systems - lasers and Infrared (IR) camera for navigation through smoke (not always).

- Other- life line, miner belt, gas detector, maps, tools, sounding stick, etc.

Elements of a typical mine rescue ensemble.

Some of the hazards faced in the mines may stemmed from team members having to spend long periods of time outside the mine, on the surface where climatic conditions may be cold winter conditions, or other seasonal conditions. In result, the range of hazards faced by team members may be broader than the mine conditions.

A typical mine rescue ensemble which is shown in figure weighs approximately 50 pounds. Mine rescue team members generally carry approximately a total of 60 pounds of additional weight, with the added equipment needed to perform their tasks (portable radio, life lines, gas detector, miscellaneous tools, and sounding stick). This added weight reduces the mobility, increases the discomfort level and may lead to fatigue. It has been found that the ensembles can trap body heat, leading to the risk of heat stress-related injuries. Similar issues are prevalent in the fire service

Permissions

All chapters in this book are published with permission under the Creative Commons Attribution Share Alike License or equivalent. Every chapter published in this book has been scrutinized by our experts. Their significance has been extensively debated. The topics covered herein carry significant information for a comprehensive understanding. They may even be implemented as practical applications or may be referred to as a beginning point for further studies.

We would like to thank the editorial team for lending their expertise to make the book truly unique. They have played a crucial role in the development of this book. Without their invaluable contributions this book wouldn't have been possible. They have made vital efforts to compile up to date information on the varied aspects of this subject to make this book a valuable addition to the collection of many professionals and students.

This book was conceptualized with the vision of imparting up-to-date and integrated information in this field. To ensure the same, a matchless editorial board was set up. Every individual on the board went through rigorous rounds of assessment to prove their worth. After which they invested a large part of their time researching and compiling the most relevant data for our readers.

The editorial board has been involved in producing this book since its inception. They have spent rigorous hours researching and exploring the diverse topics which have resulted in the successful publishing of this book. They have passed on their knowledge of decades through this book. To expedite this challenging task, the publisher supported the team at every step. A small team of assistant editors was also appointed to further simplify the editing procedure and attain best results for the readers.

Apart from the editorial board, the designing team has also invested a significant amount of their time in understanding the subject and creating the most relevant covers. They scrutinized every image to scout for the most suitable representation of the subject and create an appropriate cover for the book.

The publishing team has been an ardent support to the editorial, designing and production team. Their endless efforts to recruit the best for this project, has resulted in the accomplishment of this book. They are a veteran in the field of academics and their pool of knowledge is as vast as their experience in printing. Their expertise and guidance has proved useful at every step. Their uncompromising quality standards have made this book an exceptional effort. Their encouragement from time to time has been an inspiration for everyone.

The publisher and the editorial board hope that this book will prove to be a valuable piece of knowledge for students, practitioners and scholars across the globe.

Index